Trends in Logic

Volume 45

TRENDS IN LOGIC
Studia Logica Library

VOLUME 45

SCOPE OF THE SERIES

The book series Trends in Logic covers essentially the same areas as the journal Studia Logica, that is, contemporary formal logic and its applications and relations to other disciplines. The series aims at publishing monographs and thematically coherent volumes dealing with important developments in logic and presenting significant contributions to logical research.

The series is open to contributions devoted to topics ranging from algebraic logic, model theory, proof theory, philosophical logic, non-classical logic, and logic in computer science to mathematical linguistics and formal epistemology. However, this list is not exhaustive, moreover, the range of applications, comparisons and sources of inspiration is open and evolves over time.

More information about this series at http://www.springer.com/series/6645

Holger Andreas · Peter Verdée
Editors

Logical Studies of Paraconsistent Reasoning in Science and Mathematics

Springer

Editors
Holger Andreas
University of British Columbia
Kelowna, BC
Canada

Peter Verdée
Institut Supérieur de Philosophie
Université Catholique de Louvain
Louvain-la-Neuve
Belgium

ISSN 1572-6126 ISSN 2212-7313 (electronic)
Trends in Logic
ISBN 978-3-319-40218-5 ISBN 978-3-319-40220-8 (eBook)
DOI 10.1007/978-3-319-40220-8

Library of Congress Control Number: 2016949620

Printed on acid-free paper

This Springer imprint is published by Springer Nature
The registered company is Springer International Publishing AG
The registered company address is: Gewerbestrasse 11, 6330 Cham, Switzerland

Contents

Paraconsistent Reasoning in Science and Mathematics: Introduction

Holger Andreas and Peter Verdée

In this book we present a collection of papers on the topic of applying paraconsistent logic to solve inconsistency related problems in science, mathematics and computer science. The goal is to develop, compare, and evaluate different ways of applying paraconsistent logic. After more than 60 years of mainly theoretical developments in many independent systems of paraconsistent logic, we believe the time has come to compare and apply the developed systems in order to increase our philosophical understanding of reasoning when faced with inconsistencies. This book wants to be a first step toward an application based, constructive debate to tackle the question which systems are best applied for which kind of problems and which philosophical conclusions can be drawn from such applications.

In this introduction we begin with a short but original overview and categorization of the research area of paraconsistency. We present some often heard reasons to go paraconsistent, a number of strategies to formally obtain paraconsistency and a couple of possible objections against paraconsistency. We hope that this way also readers new to the field can find their way inside a sometimes ill-structured but very interesting debate. The goal of this overview is therefore not at all encyclopaedic or historical, but we aim to enable the reader to enter and structure the field with a problem solving attitude: what are the problems paraconsistent logicians want to solve, what are the strategies they use for solving them and what are the main difficulties in the process toward the solution?

Paraconsistency is not a well defined notion. Paraconsistent reasoning could be seen as any kind of reasoning which is able to deal with inconsistencies. Paraconsistent logics propose systematic ways to reason paraconsistently. In this introduction

H. Andreas (✉)
University of British Columbia, Kelowna, Canada
e-mail: holger.andreas@ubc.ca

P. Verdée
Université catholique de Louvain, Louvain-la-Neuve, Belgium
e-mail: peter.verdee@uclouvain.be

H. Andreas and P. Verdée (eds.), *Logical Studies of Paraconsistent Reasoning in Science and Mathematics*, Trends in Logic 45,
DOI 10.1007/978-3-319-40220-8_1

we will not make a distinction between inconsistencies and contradictions. Both terms are used to indicate information from which, for some sentence A, both A and not-A can be obtained.

The only characteristic all paraconsistent logics have in common is that the *Explosion rule*, i.e. "derive B from A and $\neg A$" where $\neg$ is a negation connective, is not valid. To people who have not studied formal logic, this rule usually comes across as very awkward. It seems always unreasonable that the derivation of arbitrary conclusions is possible, no matter what the premises are. Most people will admit that contradictions and contradictory theories are false, but deny that from false information arbitrary conclusions can be obtained. There is no consensus on what should be the rational alternative to Explosion. Some people will for example rather suggest the opposite of Explosion: forbid to derive anything further once one has obtained a contradiction. But this is not the general strategy of paraconsistent logics: most of them will still allow some more innocent consequences of contradictory theories.

Although Explosion is generally not considered as pre-theoretically valid, and is (to our knowledge) never applied in actual reasoning or informal proofs, the rule is valid in the vast majority of theoretically elaborated symbolic logics (classical logic, intuitionistic logic, fuzzy logic, their extensions - most modal, deontic, temporal logics, and many more). The reason for this contrast between the counter-intuitive character and general formal validity of Explosion is its connectedness with other much more plausible principles of reasoning. Much more intuitive are the principles of Disjunctive Syllogism (From $\neg A$ and $A \vee B$, derive B) and Addition or Disjunction Introduction (From A, derive $A \vee B$). It is impossible to unrestrictedly validate both rules in a Transitive system (in which formulas derived by means of rules can be used as premises for the application of other rules), without also implicitly validating the Explosion principle. To see this, consider that Addition enables us to derive $A \vee B$, for each arbitrary formula B, from premises A and $\neg A$. If we subsequently apply Disjunctive Syllogism to this conclusion and the second premise, we immediately obtain B, which was an arbitrary formula (possibly unrelated to the premises).

However, there is no principled reason why a symbolic logic needs to validate Explosion. For various reasons one may want a logic with exactly the property not to validate Explosion. Of course one will also lose some other properties of traditional explosive symbolic logics. We believe that one should not be dogmatic about symbolic logic. The formal theory of logic is, just like any other theory, but an attempt to capture external phenomena. Such an attempt is a fallible enterprise. Even if one believes that there is one true ultimate logic, there is no absolute warrant that our present most popular logical theories have correctly captured this ultimate logic. Even if one argues that Explosion is ultimately a valid inference, one should explain how rational agents deal with inconsistencies. If this can happen in a systematic way, there is no reason why we should not explain it by means of a logic, where logic is here understood in a maximally broad sense: as any symbolic way to theorize about reasoning. Both the reasons why one wants a paraconsistent logic and the ways in which paraconsistency is obtained may be very diverse. Let us first list some of the reasons philosophers may have to develop paraconsistency.

The first and most obvious reason is the need to deal with inconsistent information or inconsistent theories. It is rather uncontroversial that every human attempt to obtain empirical or theoretic information is susceptible to inconsistencies. The reasons for this may be very diverse. There may be errors in the processing of information, errors in storing and retrieving information, calculation errors, errors in observation, discrepancies between theory and observations, unforeseen contradictory outcomes of theories, incompatibilities between different of our best theories about the world (possibly all empirically adequate w.r.t. past observations), inconsistent sources, inconsistent databases, defeated conclusions of non-deductive reasoning...All human epistemic methods are fallible and in case of failure there is nothing which can warrant the freedom from inconsistencies. Each part of our knowledge can in principle be wrong and then be in conflict with other parts or future observations. If this happens, it is rarely easy to solve the issue by diagnosing the problem and simply contracting the problematic sentences. These conflicts are parts of a structured web full of useful information. But even if we are able to remove mistaken information, there usually is no immediate correct alternative at hand which is harmless. As long as there is no alternative, one seems to be forced to provisionally take this inconsistent body of information as it is and continue reasoning from there, until one has found a way to solve the problem.

An Explosive logic cannot be used for this purpose, because in such a logic every inconsistent theory is interderivable (from any inconsistent theory one can derive all statements of any other theory) and thus equivalent. This means that, given an Explosive logic, all inconsistent information becomes inferentially identical and therefore entirely useless. It should not come as a surprise that this is highly undesired for the above described type of problems. Instead one may want a logic that maximally isolates (possible) inconsistencies so that the underlying problems do not infect or affect other parts of our knowledge. Or, on the other side of the spectrum, one may rather want a logic which maximally approximates an Explosive logic, but without Explosion. Similarly, one may want a logic which spreads inconsistencies to all formally related sentences to avoid potentially false assumptions of safety for indirectly affected sentences. Alternatively, one may want a mechanism to "repair" the inconsistency. Yet another project may be to devise a logic which reduces inconsistencies to more basic/primitive inconsistencies.

A second reason is dealing with inescapable, acceptable or true contradictions. This concerns several versions of dialetheism (cf. [22]). Semantic paradoxes (among which the famous Liar Paradox) show that we cannot combine traditional Explosive logic, certain parts of ordinary language (e.g. unrestricted self reference) and straightforward principles of semantics (e.g. transparent truth). Other paradoxes (set theoretic, property theoretic, related to informal mathematical proofs or definitions) show that inconsistencies are obtained by using certain intuitively very attractive principles of reasoning. In all these cases ways have been found to avoid the inconsistencies by restricting the modelled domains and the validity of the intuitive principles. But one may choose to take the intuitions behind the problematic theories seriously and so to bite the bullet and accept the inescapable inconsistency. Of course one needs a logic to reason with such an acceptable or even true inconsistency. Even if one does not want to make the inconsistency true in a strong *truth as correspondence*-sense, one may see a mathematical or semantic theory in a less realistic way and, because of external reasons, argue that an inconsistent theory is preferable over its possible consistent corrections as the most appropriate theory of a certain domain.

A third reason is dealing with the possibility of inconsistent (counterfactual) worlds. Even if inconsistent objects do not exists, one may find it useful or even metaphysically required to be able to reason with them. A Meinongian, for example, who believes that inconsistent objects do not exist, may still see them as objects that we can describe in a reasonable language. If they have inconsistent properties, an Explosive logic cannot be the underlying logic of such a language. One may also reason that logic should be maximally neutral. If the logic excludes inconsistent theories already in advance, logic seriously restricts the metaphysical possibilities. So even if one is strongly convinced that there are no true inconsistencies, one may see this as a matter of fact and not as an a priori truth determined by logic.

A fourth reason may be dealing with the entailment/implication connective as used in informal mathematics or science. Independent of one's considerations about the nature of negation and inconsistencies, logicians, such as the fathers of relevance logics (cf. e.g. [1, 12]), have attempted to give a reasonable formalization of implication connectives (closer to actual usage than material implication). They took it to be essential for implication to express a link between antecedent and consequent. There is of course no link whatsoever between p-and-not-p and q, so p-and-not-p cannot imply q, and so the object language variant of Explosion $(A \wedge \neg A) \rightarrow B$ cannot be valid in a logic based on such a view on implication. In a sense then, such logics are also paraconsistent, even if one often does not define a consequence relation but merely a set of tautologies.

A fifth reason may be the discovery of ignored domains of mathematics. In the same way as the generalization of real numbers to complex numbers turned out to be a rich broadening of mathematics, also taking inconsistent theories and inconsistent models of existing theories seriously may enlarge the mathematical domain in an interesting way. Once one has a precise paraconsistent logic to deal with inconsistencies, there is no reason why a mathematical theory could not be inconsistent, as long as it is as rigorous as the theories of classical mathematics.

For all of these reasons, paraconsistent logicians have developed a plethora of different systems in the relative short history of paraconsistency. We here list some of the most prominent approaches, divided into several categories. The categories may overlap. We certainly do not aim to give a full overview, but merely a more or less original categorisation of possible approaches.

A first category contains logics in which the consistency of sentences can be expressed formally by means of a (possibly defined) unary connective. In an Explosive logic, such a symbol would be trivial, because every sentence is supposed to be consistent in such a logic. The first examples of such logics were so called Da Costa $\mathbf{C_n}$-logics (cf. [13]) in which $\neg(A \wedge \neg A)$ is interpreted as expressing the consistency of A. Later this is generalized to the class of LFI's: Logics of Formal Inconsistency (cf. [11]). This is a general framework which contains very different paraconsistent logics with a unary consistency connective. In such logics one has Explosion for consistent formulas (and not for all other formulas). The advantage of this approach is that one can understand and model classical reasoning as well as paraconsistent reasoning, depending on which formulas are involved.

A second category are the many valued logics. The idea is to allow other truth values than consistent truth and consistent falsity. An evident choice is of course the introduction of a third value indicating both true and false. Other options are: going four valued (adding a value for neither-true-nor-false) or even infinitely valued. In general Disjunctive Syllogism will not be valid, because there are non-classical truth values that make both A and $\neg A \vee B$ true without affecting the value of B. There are a lot of examples, but prominent ones are these: the three valued logic of paradox (cf. [20]), the four valued Belnap Dunn logic (cf. [14]) and, recently, paraconsistent logics based on infinitely valued fuzzy logics [15]).

A third category are systems in which negation is a modal connective. This is a diverse group, but in general one obtains paraconsistency by interpreting negation in such a way that the truth of a negated formulas is interpreted as the possible falsity of that formula. Just like it is consistent in classical modal logic that a formula is true and possibly false, there are models in which a formula and its negation are both true, given such a modal paraconsistent negation. The first and best known example is what is now known as dual intuitionistic logic (cf. [16]). In the Kripke semantics of intuitionistic logic negation is interpreted as not possibly true, were 'possibly' means provable at some future point in time. In the dual version we could read negation as possibly false or refutable at some point in time. This treatment of negation as a modal connective can be based on many other modal logics in several diverse ways of expressing the paraconsistent negation. A philosophical study of negation as a modality can be found in [8].

A fourth category are the non-adjunctive or discussive paraconsistent logics (cf. [17]). Consistencies are here possible because a sentence may be coherently held by one agent in a discussion and its negation coherently by another agent in the same discussion. From the point of view of a neutral observer of the discussion we are dealing with an inconsistency. $A \wedge \neg A$ would still explode, but there is no way to conjoin A and $\neg A$. In these logics Addition is unproblematic, but Disjunctive Syllogism is only valid in its one premise form: from $A \wedge (\neg A \vee B)$ conclude B. The Adjunction (Conjunction Introduction) rule (derive $A \wedge B$ from A and B) is blocked.

A fifth category are the non-monotonic logics. They restrict the law of Disjunctive Syllogism to formulas which could be consistent in view of the premises. For the other formulas one of the other paraconsistency strategies is used. That way only those models are selected that verify a minimal amount of inconsistencies. The advantage of this strategy is that one has the inferential richness of Explosive logics for consistent premises, but also the inconsistency tolerance when the premises are not consistent. The most well known examples are inconsistency adaptive logics [5] and minimally inconsistent **LP** (cf. [21]).

A sixth category contains logics that are not cautiously transitive. Merely blocking (cautious) transitivity makes it possible to validate both Disjunctive Syllogism and Addition (and in fact all the strength of classical logic for consistent premises) without necessarily validating the Explosion rule (cf. [6, 9]). Non-monotonic logics will usually also be non-transitive but will make sure that they are cautiously transitive (if something follows from the combination of the premises plus a conclusion, then it is also a valid conclusion of the premises alone). If one does not have cautious transitivity, one can have both unproblematic rules, as long as both are not applied in chain (one after the other). In this case one can have a monotonic structural paraconsistent logic which validates all classical consequences of consistent premises. The price to pay here is the capacity to build on earlier results. Every proof needs to start again from the basic axioms.

A seventh category are the implication revising logics. It is quite generally recognized that the material implication of classical logic (and many other logics) is far from the implication connective used in informal reasoning. A first example are the relevant logics (cf. [2] and the discussion above). Connexive logic (cf. [18]), on the other hand, are not subclassical; they really contradict classical logic. They make it false that something could imply its negation. If, by contrast, one also accepts that it is true that (some) contradictions imply their negations, one easily obtains a contradiction the logic should be able to deal with.

Still other options are logics with a non-deterministic semantics. In these systems the semantics of complex sentences are not necessarily reduced/analysed into inconsistencies concerning primitive sentences. That way the negation of a sentence A may for example be allowed to be true independent of the truth value of A. This way A and its negation may be true together. Of course one loses compositionality, but this does not need to make the logic inferentially impotent. Examples are the weaker LFIs such as **mbC** and Batens' system **CLuN** (cf. [4]).

A final category concerns the possibility to block Addition (possibly only for inconsistent formulas) in order to avoid Explosion. In a logic that merely analyses sentences into (combinations of) subsentences, of course one can never obtain an arbitrary formula. This strategy is followed in [19].

It is clear that there are a lot of paraconsistent logics with a diverse set of purposes. Although, in general, they are well-developed, both technically and philosophically, there is not much research comparing them in relation to their applicability in science and mathematics. It is surprising how little these logics are actively applied to actual scientific or mathematical theories (other than some historical reconstructions). Given how common it is that scientists have to deal with inconsistencies (between theories, between theories and observations, and inside theories) and inconsistency resolution, it is surprising that relatively little work has been done to make the involved type of reasoning logically precise.

There may be several reasons for this. Let us summarize some of the possible objections one may have against adopting a paraconsistent logic for concrete applications in science and mathematics. Many people are reluctant to use paraconsistent logics because when adopting them, one loses the strongest possible argument to reject problematic hypotheses and theories, i.e. the fact that they are inconsistent. In a paraconsistent context, logic alone does not suffice to reject inconsistent theories. Consequently, in such a context new information will never, by pure force of logic, necessitate the revision of old information. Logic no longer excludes the possibility to keep piling up all kinds of inconsistent information without ever contracting old information. Belief revision is therefore no longer a logical requirement. There may of course be many other reasonable criteria for rejecting problematic theories (incoherence, vagueness, lack of elegance, lack of explanatory power, empirical inadequacy etc.) and consistency may in a paraconsistent context still be a locally valid extra-logical requirement, or a property one may want to satisfy as much as possible. Nevertheless it is dialectically very attractive to possess a logical criterion to dismiss every theory from which one can derive an inconsistency.

Moreover, the discussion above shows that (most) paraconsistent logics need to lose some a priori attractive principles of reason. We have become so used to classical logic that all of the above considered principles (Disjunctive Syllogism, Addition, Adjunction, Monotonicity and Transitivity) seem very natural principles of logical consequence often successfully applied in informal mathematical or scientific reasoning. If such principles are no longer logically valid, one needs to explain the discrepancy between logic and practice. Are the apparently successful applications of the invalid principles mistakes, locally correct applications of a generally speaking invalid principle, or the result of a mistaken formalization? Moreover, if not merely metalogical principles (Monotonicity, Transitivity) but actual logic rules (Disjunctive Syllogism, Addition, Adjunction) are blocked, one loses inferential power to the effect that many theories become much weaker. This may be desirable for those parts of the considered theory where one is confronted with actual inconsistencies, but problematic where everything seems to behave consistently.

Logics that have more inferential power but require specific treatment for consistent parts of theories are often computationally highly complex. In order to find out whether a specific application of a rule is valid, one needs to know already whether an inconsistency is derivable from certain involved formulas in relation to the rest of the theory. Calculating this may be very difficult (cf. [25]). Suppose that, inside a given theory $\Gamma \cup \{A, \neg A \vee B\}$, one wants to apply Disjunctive Syllogism to obtain B. If the paraconsistent logic only allows this rule if A is consistent, one needs to find out whether $\neg A$ is derivable from Γ, before one is able to correctly apply this single rule. If Γ and A have some non-logical vocabulary in common and Γ is a large set or forms a complex theory, finding this out may be immensely time consuming or even undecidable. Computational complexity is not a conclusive reason to reject a logic, because the logic may be seen as merely the ideal but difficult or unreachable standard of reasoning. But in that case one should explain how human agents can at least approximately deal with the unreachable ideal reasoning standards in practice.

Yet another possible objection is the question whether formal logic is applicable in an inconsistent context. Those who do accept the importance of inconsistency tolerant reasoning may object that this type of reasoning (largely) happens extralogically (cf. [24]). Important extra-logical factors involved in dealing with inconsistencies are: the sources of information, the priority ordering of information and its sources, the goals of reasoning, social and dialectical dynamics of reasoning and arguing, and fallible diagnostic reasoning. Nobody will deny that such factors play a role in dealing with (at least some) inconsistencies, but that does not mean that one cannot also say something with logical generality based solely on the form of involved expressions.

Even those who accept that paraconsistent logics correctly formalize some phenomena, may still claim that, to the extent that they are useful, paraconsistent logics can be translated into more traditional explosive logics. Either one claims that what paraconsistent logicians would formalize as an inconsistency should actually be formalized differently (possibly with the same syntactic consequences). If an agent receives information A and information $\neg A$ from two equally reliable sources, it makes sense to formalize this as 'agent 1 believes A' and 'agent 2 believes $\neg A$', using doxastic or epistemic modalities. Even in explosive logics this pair does not explode. Similar modal solutions work for every kind of inconsistency coming from different origins (incompatible axioms, theories, observations). In case the inconsistency originates from one indissoluble (inconsistent) body of information this strategy does not work, but one could then argue that the body of information is simply unreliable and should not be used for doing further reasoning.

Another alternative is to consider the paraconsistent negation as a coherent unary connective that can be added to classical logic, but not as *the* negation. One can often use the usually classical semantics of the paraconsistent negation connective to add it to classical logic as a conservative extension. Or one can define inconsistency tolerant databases or inconsistent properties/collections inside a purely classical context as well-defined mathematical objects. Compare it to fuzzy set theory. People speak of fuzzy *sets* (cf. [26]), but they are merely useful classical mathematics tools, which are defined by means of ordinary sets and real number theory. They behave in such a way that they are more subtle generalizations of ordinary sets (ones to which the elements belong to a certain degree, instead of just in or out). They are no alternatives to classical set theory, but mere extensions of it. People who accept the usefulness of some paraconsistent logics can claim the same thing about a paraconsistent negation; a useful tool that can be defined in a rich enough classical logic (plus perhaps some parts of classical set theory). For many applications of paraconsistent logic it seems indeed unnecessary to really revise classical logic; it is often sufficient to realize that classical logic is not the appropriate tool to approach inconsistent collections of information. But classical logic was never meant for this purpose anyway. The idea would be that one could keep using classical logic with its inconsistency intolerant negation for all the more foundational/justifying purposes it was meant for. This may be a reasonable position if it concerns rather practical applications of paraconsistent logic, but it does not work for more fundamental applications about the very basics of mathematics, philosophy and logic.

We have listed a number of often heard objections to the usefulness of paraconsistent logic as an alternative to classical logic. None of these objections are sufficient arguments to reject the usefulness of paraconsistent logics, but those defending paraconsistent logics need to specify how to overcome these issues. This itself is an interesting debate and the possible answers depend a lot on which logic and which application one has in mind.

The reader understands by now that paraconsistency is a diverse phenomenon with different *raisons d'être*, different technical solution and different ways to respond to criticism, all of which have to do with the specific application one has in mind. Nevertheless there is also quite a lot of common ground. Similar techniques have been used, similar arguments have been given against Explosion and against the critics of paraconsistency, and similar inconsistent theories have been studied, all of this often independently from one another and inside different schools of paraconsistency. Nevertheless there is relatively little study about the similarities and differences of the different currents of paraconsistency in relation to the intended real life applications of paraconsistency.

In order to open the debate on how the different formalisms relate to their real life applications in the philosophy of science and mathematics, we decided to organize a conference on this topic in Munich, Germany: the conference *Paraconsistent Reasoning in Science and Mathematics* (June 11–13, 2014) in the beautiful setting of the Carl-Friedrich-von-Siemens-Stiftung. Our aim was to bring the different schools of paraconsistency together to open the debate on how the different formalisms relate to their real life applications in the philosophy of science and mathematics.

The level of the talks and the quality of the debate was so high that the participants of the conference were all in favour of publishing a volume on the topic of the conference, aiming toward a written and more detailed follow up of this debate. The present book is the result. We hope the reader will find that it lives up to the expectations. In what follows, we give a brief summary of every paper of our collection.

1 Holger Andreas and Peter Verdée: Adaptive Proofs for Networks of Partial Structures

According to Carnap [10], we interpret and understand the theoretical terms of a theory T in such a manner that the axioms of T come out true. If, however, T is classically inconsistent, this semantic doctrine fails to work properly. The theoretical terms remain uninterpreted in this case. This is not satisfactory insofar as numerous scientific theories turned out to be inconsistent in some way or other – science is full of inconsistencies. Hence, it is desirable to have a semantics of theoretical terms that also applies to inconsistent theories.

Holger Andreas and Peter Verdée, consequently, develop a paraconsistent generalization of the semantic doctrine in question: we interpret and understand the theoretical terms of a theory T in such a manner that the axioms of T are satisfied to a maximal extent. Formally, we describe such interpretations in terms of a network of partial structures, and thereby define a preferred models semantics of paraconsistent scientific reasoning. This semantics respectively defines an inference relation for flat and prioritized axiomatic theories.

A preferred models semantics by itself does not give us a proof-theoretic account of scientific reasoning with theoretical terms. For this to be achieved, the framework of adaptive logic with its dynamic proof-theory suggests itself. Hence, we present a flat and a lexicographic adaptive logic which are proven to capture the inference relation for flat and prioritized axiomatic theories. Because the adaptive logics belong to the category of standard (lexicographic) adaptive logics, the adaptive characterization immediately gives rise to an adequate dynamic proof theory for the inference relations. The paper concludes with a demonstration of how we can derive sensible conclusions from Bohr's model of the atom using adaptive proofs.

2 Franzesco Berto: Ceteris Paribus Imagination

Franzesco Berto explores impossible worlds for an analysis of ceteris paribus imagination. An impossible world is one where the laws of classical logic may be violated by the truth-value assignment to atomic and complex formulas. Hence, an impossible world may verify a set of sentences that is classically inconsistent. Impossible worlds, therefore, may serve as a model of inconsistent beliefs.

Why should we want to model inconsistent beliefs? The underlying motivation derives from the limitations of our logical capacities. We are unable to grasp all logical consequences of a set of explicit beliefs, and we may even fail to recognize inconsistencies in our explicit beliefs. As is well known, this happened to Frege when he developed his *Basic Laws of Arithmetic*. In brief, we are not logically omniscient.

Ceteris paribus imagination is modelled by a conditional: if an agent explicitly conceives A to be the case, then B is part of the imagined scenario. In formal terms: $[A]B$, where $[*]$ is a modal operator, defined by an accessibility relation on the set of possible and impossible worlds. [A] B holds true if B is verified by all worlds (possible and impossible ones) that are reachable from the actual world and in which A holds true.

Having defined a worlds semantics of $[A]B$, Berto investigates which axioms envisioned for variably strict conditionals remain valid for ceteris paribus imagination. Notably, $[A \wedge \neg A]B$ fails to hold for arbitrary B. Imagining an inconsistent scenario does not mean that we trivialize what we conceive. In this respect, the ceteris paribus conditional behaves like a paraconsistent consequence relation.

3 Bryson Brown: On the Preservation of Reliability

Science is full of inconsistencies: first, we have scientific theories that are internally inconsistent and thus imply a contradiction. Second, we have scientific theories that make assumptions inconsistent with other accepted scientific theories. Third, we have numerous approximations and idealizations that are known to be inconsistent with what we strictly believe about the respective theoretical entities. Fourth, scientific theories are often times inconsistent with certain predecessor theories, while preserving many of their empirical predictions. These inconsistencies strongly suggest the need for a paraconsistent treatment of scientific reasoning.

Bryson Brown attempts to provide methodological foundations for a paraconsistent approach to scientific reasoning. His proposal is to view reliability-in place of truth-as the property to be preserved by proper scientific reasoning, as well as in the replacement of earlier scientific theories by new ones. The main focus of Brown's paper is on reliable inference patterns in the history of science, including Planck's treatment of black body radiation and Bohr's theory of the hydrogen atom; work by Nancy Cartwright and Bas C. van Fraassen is also discussed, leading up to an account of a modestly paraconsistent approach to scientific reasoning.

4 Luis Estrada-González: Prospects for Triviality

This paper studies triviality in mathematical theories, an important enemy of most paraconsistent logicians. Paraconsistent logics (want to) serve as the underlying logic of inconsistent theories, exactly because they can avoid triviality. Trivial theories are usually seen as meaningless and even disastrous. This position was among others defended by Chris Mortensen.

The author of this paper discusses the central question whether triviality is always so bad and wants to answer it in the negative, against Mortensen's position. He argues that there is a case of an extremely simple mathematical category theoretic universe, a degenerate topos, in which everything is true. This universe is therefore trivial, but it is not inherently problematic.

Mortensen's case is built on a trivialisation result for real number theory. González shows that either one of the premises of the trivialization result cannot obtain (from a point of view external to the universe) and thus the argument is unsound, or that it obtains in calculations internal to such a trivial universe. In the latter case the calculations in the trivial universe are possible and meaningful albeit extremely simple. Our actual universe is probably not as simple as and so does not correspond to this degenerate topos, but that does not mean that what is done inside the universe is meaningless.

5 Andreas Kapsner: On Gluts in Mathematics and Science

This paper discusses the question whether truth value gluts (both true and false) should be designated in an analysis of mathematical and scientific reasoning. Practically speaking the question is whether one should assert sentences that are true and false and whether they should be used as basis for decisions and actions and as premises of arguments. The traditional paraconsistent view is that there are truth value gluts and that they should be designated. In some sense the converse goes against the very basic starting point of paraconsistency: a non-designated glut will not block the Explosion rules.

Kapsner defends the view that it is often, but not always, unreasonable to assert glutted statements. He presents a clear case: if two costumer reviews contradict each other on the quality of a product, one should not assert the contradicting information obtained by reading the reviews. Subsequently he presents some cases from the history of science (the infinitesimal calculus and the Darwin–Kelvin debate on the age of the earth) to indicate that also in these case it may be more reasonable not to designate gluts.

6 Carlos A. Oller: Contradictoriness, Paraconsistent Negation and Non-intended Models of Classical Logic

This paper concerns the often heard argument that paraconsistent negation is not a real negation because a sentence and its negation should never be true together. The author attacks the argument by showing that it could also be used to show that classical logic's negation is by the same standards not a real negation either.

Classical logic has certain unavoidable non-intended models. Carnap was the first to point out that adding a trivial model (in which all formulas are true) to the semantics of classical logic does not affect the set of valid consequences. In such a model of course formulas and their negations are both true. It seems thus that it is impossible even in classical logic to exclude the possibility that a formula and its negation are both true.

7 Hitoshi Omori: From Paraconsistent Logic to Dialetheic Logic

This paper proposes a new approach to paraconsistent logic to be applied as the underlying logic of a dialetheic version of naive set theory and naive truth. The author proposes an attractive logic which is not only paraconsistent in that it can tolerate inconsistent formulas, but also dialetheic, in the sense that it also makes certain inconsistent formulas into tautologies.

Omori returns to the modern origins of paraconsistent logic and proposes a paraconsistent account of negation in line with some ideas by Jaśkowski: that a good negation should be a connective such that each formula and its negation form contradictory pairs. This is realized by requiring that a formula is true iff its negation is false, and false iff its negation is true. A necessary condition for paraconsistent logics respecting this account of negation is that Double Negation rules are valid.

In order for the logic to work with prototypical inconsistent mathematical theories such as naive set theory, one needs a weak enough biconditional to non-trivially express axiom schemas like the axiom of Abstraction. To this purpose a strategy suggested by Laura Goodship is used: opt for the material biconditional of **LP**. Omori adds to this concept some ingredients of LFIs (a consistency operator) and connexive logics (the conditional is false when the antecedent is not true or the consequent is false). The result is a dialetheic logic with a functionally complete three valued semantics.

8 Martin Pleitz: Paradoxes of Expression

The history of the paradoxes and attempted solutions thereof shows many cases where a certain solution falls prey to another, more refined variant of the original paradox. The revenge liar is the most famous instance of such cases. Martin Pleitz adds another chapter to this history of attempted solutions and recurrent paradoxes.

The focus is on a recent proposal by Graham Priest to solve the semantic and set-theoretic paradoxes using a biconditional that does not detach, i.e., that fails to satisfy *modus ponens*. A detachable truth schema, however, is needed for what has been described as *blind endorsement*. For example, if one holds that everything that the Bible says is true, one blindly endorses all the claims made in the Bible. To solve this problem, Priest entertains the idea of introducing a detachable conditional, together with an expression predicate that allows us to say that certain propositions are expressed by certain sentences.

Martin Pleitz formulates a very reasonable principle that an expression operator, licensing blind endorsements, should satisfy: any meaningful sentence should be synonymous with itself. Based on an axiomatic formulation of this principle, he shows that variants of the Liar and the Curry paradox can be formulated. Hence, we have a contradiction and a way to trivialize the system envisioned by Priest. As triviality is unbearable even in a paraconsistent setting, this casts serious doubt on Priest's proposed solution.

9 Corry Shores: Dialetheism in the Structure of Phenomenal Time

The very idea of motion seems to be contradictory: if we say that an object is moving, we imply that it is at different places at different times. So far, things are consistent. If, however, we want to say that an object is moving *right now*, we seem to ascribe the property of changing positions to a specific time point. At a specific time point, however, an object can only be at one place. Drawing on Zeno's paradox of the arrow and assuming that only the present time point has reality, we can thus argue that no object is really moving. Motion is not part of reality. Likewise, the flow of time is an unreal phenomenon.

Alternatively, we can accept that motion and time are contradictory but real, thereby embracing some form dialetheism. This alternative is investigated and sympathetically entertained by Corry Shores in his contribution. Besides the work of Zeno, he draws on Husserl's phenomenology and subsequent phenomenological research to motivate a dialetheist account of change and time. Dialetheist ideas about the phenomenology of time are thus brought together with recent work in theoretical psychology.

10 Fenner S. Tanswell: Saving Proof from Paradox: Gödel's Paradox and the Inconsistency of Informal Mathematics

In this paper two of the most popular arguments (by Beall resp. Priest) in favour of the inconsistency of (informal) mathematics (and so the need to formalize it with a paraconsistent logic) are discussed. A first argument is based on what is sometimes called Gödel's paradox, i.e. a sentence expressing that it is not provable. Accepting the existence of such a sentence leads to a contradiction in mathematics. The second argument is based on the incompatibility of completeness and consistency established by Gödel's incompleteness theorems. Arguing in favour of the completeness of informal mathematics, thus also forms an argument against the consistency of mathematics.

Tanswell argues against these arguments that the necessary distinctions between formality and informality are often ignored. The author also points at problems with the assumption of the unity of informal mathematics.

11 Heinrich Wansing and Sergei Odintsov: On the Methodology of Paraconsistent Logic

When we decided to organize our conference on *Paraconsistent Reasoning in Science and Mathematics*, we wanted to stimulate discussion, exchange of ideas, and further research on the desiderata that a paraconsistent logic should satisfy. We argued there to be three core desiderata: (1) A paraconsistent logic ought to capture the inferential use of inconsistent but non-trivial theories. (2) A paraconsistent approach should explain how one can weaken the underlying logic of classical logic to get rid of the explosion principle and still have enough inferential power to be successful. (3) It is desirable to have a philosophical motivation for the deviation from classical logic in terms of epistemological and, possibly, also metaphysical considerations.

We have then been very pleased to see that Heinrich Wansing and Sergei Odintsov directly address the question of which desiderata a paraconsistent logic should satisfy. While investigating the historical roots of the above desiderata, they cast some doubt on desiderata (2) and (3). More precisely, they question that the reference logic of a paraconsistent logic should be classical logic, arguing that the choice of classical logic as reference logic is at least difficult to justify. As for the philosophical motivation for developing a specific paraconsistent logic, the notion of information should play a central role rather than epistemological and metaphysical considerations. As information about whatever domain is rarely complete and often times inconsistent, an informational methodology of paraconsistent logic may lead us to choosing a non-bivalent logic as reference logic.

The paper further discusses in great detail the maximality condition that a paraconsistent logic should satisfy with reference to classical logic, thus drawing on work by Arieli et al. [3]. Moreover, it examines methodological considerations on the desiderata of a paraconsistent logic that have been suggested by Priest and Routley [23]. Finally, Wansing and Odintsov sketch a universal approach to constructing a paraconsistent logic for a given reference logic that may well not be classical logic.

12 Zach Weber: Paraconsistent Computation and Dialetheic Machines

This paper concerns the application of paraconsistent logic and dialetheism to theoretic computer science. The question is asked whether there are algorithms which are essentially paraconsistent, in the sense that only paraconsistent logic can recognize them. While the question may seem counterintuitive, it is clear that certain objects can exists in paraconsistent mathematics which cannot exist otherwise (for example the Russell set or the set of all ordinals). So it is not unlikely that also the concept of an algorithm should be reconsidered in a paraconsistent setting in order for classically unknowable but useful objects to be recognized and studied.

The author argues in favour of the existence of such properly paraconsistent computations. Arguments by Sylvan and Copeland, Routley, and Priest support this view. One of the arguments goes as follows: in view of a straightforward diagonalization, the algorithm that enumerates all algorithms (intuitively) is but at the same time cannot be an algorithm. If it is an algorithm (and it sure seems to be one) it has to be an inconsistent algorithm.

Subsequently Weber investigates the ways in which one could formulate paraconsistent algorithms in a dialetheic mathematical metalanguage. He discusses the properties of so called dialetheic machines and their relation with finiteness and the halting problem.

References

1. Ackermann, W. (1956). Begründung einer strengen Implikation. *Journal of Symbolic Logic*, *21*, 113–128.
2. Anderson, A. R., & Jr Belnap, N. D. (1975). *Entailment. The logic of relevance and necessity* (Vol. 1). Princeton: Princeton University Press.
3. Arieli, O., Avron, A., & Zamansky, A. (2011). Ideal paraconsistent logics. *Studia Logica*, *99*, 31–60.
4. Batens, D. (1999). Inconsistency-adaptive logics. In E. Orłowska (Ed.), *Logic at work. Essays dedicated to the memory of Helena Rasiowa* (pp. 445–472). Heidelberg, New York: Physica Verlag (Springer).
5. Batens, D. (2000). A survey of inconsistency-adaptive logics. In [7] (pp. 49–73).
6. Batens, D. (2012). It might have been classical logic. *Logique et Analyse*, *55*(218), 241.
7. Batens, D., Mortensen, C., Priest, G., & Van Bendegem, J. P. (Eds.). (2000). *Frontiers of paraconsistent logic*. Baldock, UK: Research Studies Press.
8. Berto, F. (2015). A modality called 'negation'. *Mind*, *124*(495), 761–793. doi:10.1093/mind/fzv026.

9. Besnard, P., & Hunter, A. (1995). Quasi-classical logic: Non-trivializable classical reasoning from inconsistent information. *Symbolic and quantitative approaches to reasoning and uncertainty* (pp. 44–51). Heidelberg: Springer.
10. Carnap, R. (1958). Beobachtungssprache und theoretische Sprache. *Dialectica, 12*, 236–248.
11. Carnielli, W. A., Coniglio, M. E., & Marcos, J. (2007). Logics of formal inconsistency. In D. Gabbay & F. Guenthner (Eds.), *Handbook of philosophical logic* (Vol. 14, pp. 1–93). Heidelberg: Springer.
12. Church, A. (1951). The weak theory of implication. In: A. Menne, A. Wilhemy & H. Angsil (Eds.), *Kontrolliertes Denken. Untersuchungen zum Logikkalkül und zur Logik der Einzelwissenschaften* (pp. 22–37). Karl Alber, München. (abstract: *JSL*, 16, 1951, p. 239).
13. da Costa, N. C. (1963). Calculs propositionnels pour les systèmes formels inconsistants. *Comptes rendus de l'Académie des sciences de Paris, 259*, 3790–3792.
14. Dunn, J. M. (1976). Intuitive semantics for first-degree entailments and 'coupled trees'. *Philosophical Studies, 29*(3), 149–168.
15. Ertola, R., Esteva, F., Flaminio, T., Godo, L., & Noguera, C. (2015). Paraconsistency properties in degree-preserving fuzzy logics. *Soft Computing, 19*(3), 531–546.
16. Goodman, N. D. (1981). The logic of contradiction. *Mathematical Logic Quarterly, 27*(8–10), 119–126. doi:10.1002/malq.19810270803.
17. Jaśkowski, S. (1969). Propositional calculus for contradictory deductive systems. *Studia Logica, 24*, 243–257.
18. Mccall, S. (1966). Connexive implication. *The Journal of Symbolic Logic, 31*, 415–433. doi:10.2307/2270458. http://journals.cambridge.org/article_S0022481200066822.
19. Meheus, J. (2000). An extremely rich paraconsistent logic and the adaptive logic based on it. In [7] (pp. 189–201).
20. Priest, G. (1979). The logic of paradox. *Journal of Philosophical Logic, 8*, 219–241.
21. Priest, G. (1991). Minimally inconsistent LP. *Studia Logica, 50*, 321–331.
22. Priest, G. (2006). *In contradiction. A study of the transconsistent* (2nd ed.). Oxford: Oxford University Press. (first edition 1987).
23. Priest, G., & Routley, R. (1989). Systems of paraconsistent logic. In F. Priest, R. Routley, & J. Norman (Eds.), *Paraconsistent logic: Essays on the inconsistent* (pp. 142–155). Munich: Philosophia Verlag.
24. Urbaniak, R., & Siniło, P. (2014). The inapplicability of (selected) paraconsistent logics. *Journal of Applied Non-Classical Logics, 24*(4), 368–383.
25. Verdée, P. (2009). Adaptive logics using the minimal abnormality strategy are Π^1_1-complex. *Synthese, 167*, 93–104.
26. Zadeh, L. (1965). Fuzzy sets. Information. *Control, 8*, 338–353.

Adaptive Proofs for Networks of Partial Structures

Holger Andreas and Peter Verdée

Abstract The present paper expounds a preferred models semantics of paraconsistent reasoning. The basic idea of this semantics is that we interpret the language $L(V)$ of a theory T in such a way that the axioms of T are satisfied to a maximal extent. These preferred interpretations are described in terms of a network of partial structures. Upon this semantic analysis of paraconsistent reasoning we develop a corresponding proof theory using adaptive logics.

Keywords Adaptive logics · Preferred models semantics · Paraconsistent reasoning · Structuralist approach

1 How to Reason with Inconsistent Theories?

As is well known, some axiomatic theories remain in use despite the observation of classical inconsistencies. Axiomatic theories of truth and naive set theory are prominent examples. Furthermore, there are well established axiomatic theories in the natural sciences that are not fully consistent with the empirical data or not consistent with certain other well established theories. The postulates of Bohr's atomic theory, for example, are not consistent with the set of Maxwell equations. Moreover, we have internally inconsistent theories in the natural sciences, such as classical electrodynamics.[1] When scientists observe such inconsistencies, they do not always abandon the scientific theory in question.

How do we reason with such inconsistent theories? Various logics and inference systems have been devised to answer this question. The present approach builds upon

[1]For a detailed investigation of the inconsistency of classical electrodynamics, see [8].

H. Andreas (✉)
University of British Columbia, Kelowna, Canada
e-mail: holger.andreas@ubc.ca

P. Verdée
Université catholique de Louvain, Louvain-la-Neuve, Belgium
e-mail: peter.verdee@uclouvain.be

H. Andreas and P. Verdée (eds.), *Logical Studies of Paraconsistent Reasoning in Science and Mathematics*, Trends in Logic 45,
DOI 10.1007/978-3-319-40220-8_2

a proposal by Rescher and Manor [13]. In essence, their analysis of paraconsistent reasoning consists of two steps:

(1) *Consolidation*, by means of which we obtain classically consistent subsets of a given, possibly inconsistent set of premises.[2]
(2) Classical reasoning is then used to draw inferences from these consolidated premise sets.

This approach can be motivated by the methodological doctrine *ex contradictione nihil sequitur*.[3] Contradictions do not have any consequences. Hence, we need to first consolidate our premises – in the sense of removing inconsistencies – before we can draw any inferences.

How does consolidation work? [13] describe this operation in terms of maximal consistent subsets of a given set of premises. A set S' is a maximal consistent subset of a set S iff (i) $S' \subseteq S$, (ii) S is classically consistent, and (iii) there is no classically consistent set S'' such that $S' \subset S'' \subseteq S$.[4]

If we strictly apply this consolidation strategy to axiomatic systems in science and mathematics, it would not yield interesting results. For if a given axiom or axiom scheme α implies an inconsistency, α is not a member of any maximal consistent set of premises. Hence, it could not be used for scientific reasoning. The case where some axiom turns out inconsistent with a set of other axioms is more involved and will be dealt with in detail in Sect. 2.8. In essence, the problem is that if we restrict our premises to maximal consistent subsets of an inconsistent axiomatic system, then some axiom cannot be used to draw inferences in a determinate way, i.e., without at the same time drawing contradictory inferences. This does not accord with how scientists actually use inconsistent axiomatic systems. For if an inconsistent axiomatic theory remains in use, then so do *all* of its axioms. For example, Maxwell's equations and Lorentz' force law have remained in use despite the fact that they imply an inconsistency when used to determine the self-force and the self-field of the electron.[5]

To analyse scientific reasoning from inconsistent theories, we suggest taking a closer look at the universal quantifier—at occurrences where it is used to express universal validity. Our main thesis is that we do use classically inconsistent axiomatic theories by selectively accepting the instances of universal axioms. That is, for a given universal axiom α and a scientific theory T, only a subset of the whole set of instances of α may be accepted. We thus refrain from determining the self-force of the electron by Maxwell's equations and Lorentz' force law, while continuing to use these axioms in other applications of electrodynamics. Likewise, we refrain from applying certain Maxwell equations to an electron orbiting around a proton, while we continue to apply these equations to other accelerated electrons.

---

[2]The term *consolidation* is borrowed from belief revision theory (see, e.g., [9].).

[3]Cf. [17]. For a critical discussion of this doctrine, see H. Wansing et al.: *On the methodology of paraconsistent logic* (this volume).

[4]See [6] for a related strategy of dealing with inconsistencies.

[5]See [8, Chap. 2] for a detailed exposition of this inconsistency.

The challenge arising here is to differentiate between sound instances of an axiom (which we continue to use) and inconsistent ones (which we reject). This is a challenge because the inconsistencies we encounter in science do not always have the form of a single instance of an axiom (or axiom scheme) that is by itself inconsistent. In the case of the self-force of the electron, it is rather a set of several axioms that, if jointly applied to an electron, imply an inconsistency. Also, it is desirable to account for scientific reasoning with axiomatic theories that fail to be empirically adequate, and hence are not fully consistent with the empirical facts, while still being used.

We describe the demarcation between sound and inconsistent instances of universal axioms in terms of a *preferred models semantics*. Using such a semantics, we define an operation of consolidation in the following way. First, we define the set of *preferred interpretations* of the language $L(V)$ of a theory T (where V stands for the descriptive vocabulary of T): an interpretation of the language $L(V)$ is preferred iff it satisfies the instances of the axioms of T to a maximal extent. Then, we can say that the consolidation of T – understood as a syntactic entity – refers to the set of those instances of T's axioms that are verified by all preferred interpretations. Our preferred models semantics, furthermore, admits classical reasoning for the consolidated parts of T.

We expound the preferred models semantics in the first part of the paper.[6] In the second part, we then develop a proof theory for this semantics using the framework of adaptive logics. The set of adaptive logics contains a wide range of logics that select interpretations of premises in such a way that a formal property of the interpretations is maximally satisfied. Because the inference relation that corresponds to the preferred models semantics selects those interpretations that verify a maximal number of instances of axioms, this inference relation can indeed be characterized by an adaptive logic (within an existing generic format of adaptive logics—the lexicographic format). The purpose of characterizing an existing inference relation by means of an adaptive logic in lexicographic format is that one immediately obtains a sound and complete dynamic proof system for the inference relation. These dynamic adaptive proofs have been developed to model the reasoning processes human agents use to reason towards ultimately correct inferences in a defeasible 'hit and miss'-like way.

2 Networks of Partial Structures

2.1 Partial Structures and Their Extensions

We shall use partial structures to represent the semantics of the instances of universal axioms. This notion is adopted from the framework of partial structures and partial

[6]This part is based on [1], which develops the network formalism as a paraconsistent semantics of theoretical terms.

truth as expounded in [7]. A partial structure is a set-theoretic structure of the form[7]:

$$\mathcal{A} = \langle A, R_k \rangle_{k \in K},$$

where A is the domain of interpretation, R_k are partial relations, and K an index set. $\mathcal{A}$ thus encodes a partial interpretation of some language L. Partiality of an n-ary relation R_k is to be understood as follows. Suppose we have an n-tuple $\langle a_1, \ldots, a_n \rangle$ (where $a_1, \ldots, a_n \in A$) such that $\langle a_1, \ldots, a_n \rangle \notin R_k$. On the semantics of partial truth, this does not imply that $R_k(a_1, \ldots, a_n)$ is false.

Partiality of a relation is more precisely accounted for by distinguishing between the positive extension R_k^+, the negative extension R_k^-, and the "neutral" extension R_k^0. For simplicity, we assume that $R_k^+ = R_k$ and $R_k^- = \emptyset$. On these two assumptions, there is no need to notationally distinguish between the positive, the negative, and the neutral extension of a relation symbol R_k in a partial structure. This has the consequence that $R_k(a_1, \ldots, a_n)$ being false is not expressible by a partial structure.

Extensions of partial structures are understood in the standard way:

Definition 1 (*Extension*) Let $\mathcal{A} = \langle A, R_k \rangle_{k \in K}$ be a partial structure. Let $(R_k)_{\mathcal{A}}$ denote the relation R_k of the partial structure $\mathcal{A}$. A structure $\mathcal{B} = \langle A, R_k \rangle_{k \in K}$ of L is an extension of $\mathcal{A}$ iff for all $k \in K$, $(R_k)_{\mathcal{A}} \subseteq (R_k)_{\mathcal{B}}$.

Da Costa and French [7] also consider structures with two domains of interpretation, where one domain contains observable and the other unobservable entities. Such distinctions between different domains of interpretation can easily be introduced without requiring substantial modifications of the subsequent definitions and explanations.

2.2 Instances and Applications of Axioms

By an instance of an axiom α we mean a closed formula where all universally quantified variables—that express the universal validity of α—have been replaced by a constant. We assume that all axioms come in a standard logical format and are preceded by at least one universal quantifier. Suppose axiom α has the logical form $\forall x_1, \ldots, \forall x_n \alpha_o(x_1, \ldots, x_n)$, where α_o does not start with $\forall$. Then, any formula $\alpha_o(c_1/x_1, \ldots, c_n/x_n)$ is an instance of the axiom α, where $c_1, \ldots, c_n$ are constants. We understand the notion of an instance of an axiom scheme in the standard way: instances of axiom schemes are obtained by replacing a schematic letter of the scheme by an appropriate expression of the formal language $L(V)$.

The notion of an instance thus understood is a syntactic notion. Instances of axioms have semantic counterparts insofar as the interpreted symbols of such instances refer to certain objects with certain properties. For example, an instance of Newton's law

[7]Unlike a *simple pragmatic structure* in [7], a partial structure does not contain a set P of sentences that are taken to be true in the correspondence sense.

of gravitation is given by a pair of bodies, both of which have a distinct place in space. Such semantic counterparts of the instances of universal axioms can be represented by set-theoretic structures, as has been shown in the work of the structuralist school [2, 15]. We adopt this idea and shall assume that any instance of a universal axiom has a corresponding partial structure

$$\mathcal{A} = \langle A_1, A_2, R_k \rangle_{k \in K},$$

A_1 is the domain of entities to which α is applied. A_2 may contain mathematical entities that are needed to express the properties of the objects of A_1. R_k are relations that partially interpret the descriptive symbols of α. We assume only a partial interpretation of these symbols because the theoretical terms of α may not be completely interpreted. If the axioms of the non-formalized theory contain functions, these may be represented by relations. In the case of Newton's law of gravitation, $R_k (k \in K)$ represent position, mass, and force for the objects of a two-body system.

Partial structures that are the semantic counterparts of the instances of a universal axiom α are called *applications* of α. Being the semantic counterpart of the instance α of a universal axiom means two things. First, the domain A of $\mathcal{A}$ comprises all the entities being referred to by the constants of α. Second, the relations of $\mathcal{A}$ are partial interpretations of the relation symbols of α.

The notion of an application of an axiom is adopted from [2, 15]. It should be noted, however, that intended applications in the structuralist framework often have a more complex structure as they may comprise sets of applications of a universal axiom. For example, they may involve applications of Newton's law of gravitation for a whole period of time. For simplicity and conformity to first order logic, we assume that applications of an axiom α always correspond to an instance of α. An application in the present framework need not be intended in the sense that some scientists think or should think that an axiom applies to it.

2.3 *Modular Semantics*

The core idea of the networks formalism may be described in terms of a modular semantics. Such a semantics is obtained by two operations upon the standard semantics: (i) the descriptive vocabulary V of the axiomatic theory T is divided into subvocabularies according to the axioms of T, and (ii) these subvocabularies in turn are interpreted by partial structures that represent applications of the corresponding axiom. Let $\alpha_1, \ldots, \alpha_n$ be the axioms of T with corresponding subvocabularies $V(\alpha_1), \ldots, V(\alpha_n)$. The partial structure $\mathcal{A}_{i,j}$ represents the application j of the axiom α_i. We assume that there can only be countably many partial structures, a constraint that should be acceptable since in a finitary language there can only be countably many instances of a universal axiom. This being said, we can graphically illustrate the basic idea of a modular semantics for axiomatic theories as follows:

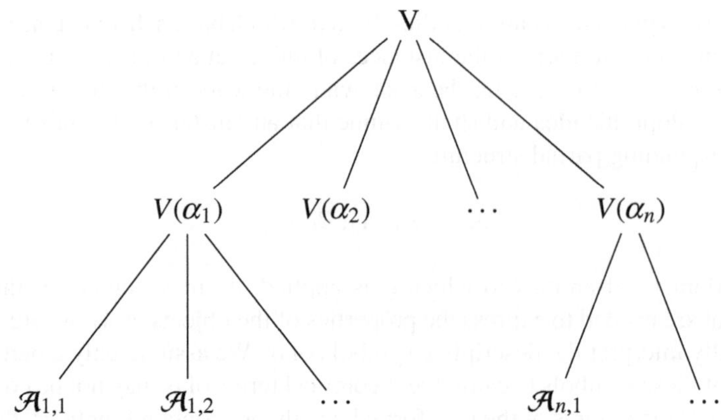

Fig. 1 Modular semantics

This modularization of the interpretation of an axiomatic theory will serve as a semantic foundation for using the instances of universal axioms selectively. Moreover, it enables us to recognize an ordering of interpretations of $L(V)$ upon which we can identify those interpretations that maximally satisfy T (Fig. 1).

2.4 Local Worlds

Let us now go further into the details of a modular semantics for axiomatic theories. Each axiom has a set of applications to empirical or abstract systems of entities. Each application of an axiom is represented by a partial structure $\mathcal{A}_{i,j}$, where i indicates the axiom α_i and j the particular application of that axiom:

$$\mathcal{A}_{i,j} = \langle A_j, A_m, R_1, \ldots, R_k \rangle.$$

The second domain A_m is introduced for properties that are expressed by mathematical objects. This domain is optional.

The structuralists describe the result of applying an axiom α to a system of entities in terms of constraints upon the (model-theoretic) extensions of a given intended application, i.e., a partial structure.[8] We adopt this idea by defining a set of "local worlds" of an application:

$$W(\mathcal{A}_{i,j}) =_{df} Mod(\alpha_i, (\mathcal{A}_{i,j})_1) \cap Ext(\mathcal{A}_{i,j}), \tag{1}$$

[8] Partial structures are used here as a generalization of intended applications. An intended application of a theory-element **T** leaves the **T**-theoretical terms undetermined. This can be represented as follows: if R_k is **T**-theoretical, $R_k = \emptyset$. Partial structures, however, are a bit more flexible as they allow us to have a partial determination of the **T**-theoretical terms for a given intended application.

where $Mod(\alpha_i, (\mathcal{A}_{i,j})_1)$ stands for the models of α_i that interpret the descriptive vocabulary of α_i in the domain of the partial structure $\mathcal{A}_{i,j}$. $Ext(\mathcal{A}_{i,j})$ denotes the (model-theoretic) extensions of $\mathcal{A}_{i,j}$. The set $W(\mathcal{A}_{i,j})$ thus is the set of extensions of the application $\mathcal{A}_{i,j}$ that satisfy the axiom α_i.

2.5 *Preferred Global Worlds*

The next step is to define the set of preferred interpretations of the language $L(V)$. We call such interpretations *preferred global worlds*. A global world – i.e., an interpretation of $L(V)$ – is preferred iff it satisfies the axioms of T to a maximal extent. A global world $\mathbf{w}$, in turn, satisfies T to a maximal extent iff there is no other global world $\mathbf{w}'$ that agrees with more applications than $\mathbf{w}$ does. This is the basic idea of what follows.

We say that a global world $\mathbf{w}$ agrees with an application $\mathcal{A}_{i,j}$ iff $\mathbf{w}$ is such that α_i actually applies to $\mathcal{A}_{i,j}$. In more technical terms, $\mathbf{w}$ agrees with an application $\mathcal{A}_{i,j}$ iff $\mathbf{w}$ contains some local world $w \in W(\mathcal{A}_{i,j})$ as a substructure. Thereupon we define a satisfaction ordering among the $L(V)$ interpretations, which determines the set of preferred global worlds. Let us begin with the relation of a substructure[9]:

Definition 2 ($Sub(w, w')$) Let $w = \langle A_1, A_2, R_k\rangle_{k\in K}$ and $w' = \langle A_1', A_2', R_k\rangle_{k\in K'}$ be two structures, or worlds. w is a substructure of w' – in symbols: $Sub(w, w')$ – iff

(1) $K \subseteq K'$
(2) $A_1 \subseteq A_1'$
(3) $A_2 \subseteq A_2'$
(4) for all $k \in K$, $(R_k)_w \subseteq (R_k)_{w'}$.

As an auxiliary notion, we define the set of subworlds of a global world:

$$Subw(\mathbf{w}) =_{df} \{w \mid Sub(w, \mathbf{w})\}. \tag{2}$$

This auxiliary notion enables us to define the set of applications with which a given global world agrees:

$$App(\mathbf{w}) =_{df} \{\mathcal{A}_{i,j} \mid \emptyset \neq W(\mathcal{A}_{i,j}) \cap Subw(\mathbf{w})\}. \tag{3}$$

The definition of the satisfaction ordering is now straightforward:

Definition 3 (*Satisfaction ordering* $<_s$) Let $\mathbf{w}$ and $\mathbf{w}'$ be interpretations of the global language $L(V)$. $\mathbf{w} <_s \mathbf{w}'$ designates the relation that $\mathbf{w}$ satisfies T to a higher extent than $\mathbf{w}'$.

$$\mathbf{w} <_s \mathbf{w}' =_{df} App(\mathbf{w}') \subset App(\mathbf{w}) \tag{4}$$

---

[9]This understanding generalizes the standard definition of being a substructure. Unlike the standard definition, it is not required that a substructure and its superstructure share the same slots of relations.

In brief, $\mathbf{w} <_s \mathbf{w}'$ iff $\mathbf{w}$ agrees with more applications of the axioms of T than $\mathbf{w}'$ does. Global worlds $\mathbf{w}$ being minimal under $<_s$ are preferred in the sense of satisfying the axioms of T to a maximal extent.

2.6 The Inference Relation

A semantics of preferred models was first introduced by Shoham [14], which was then further refined and investigated by Kraus et al. [10]. Drawing on their work, we can define an inference relation for our modular semantics. To this end, we need to introduce the notions of an $<$-minimal element and smoothness:

Definition 4 ($<$*-minimal elements*) Let $A \subseteq D$ and $<$ a binary relation on D. x is $<$-minimal in A iff there is no $x' \in D$ such that $x' < x$.

Definition 5 (*Smoothness*) Let $A \subseteq D$ and $<$ a binary relation on D. $<$ is smooth in A iff for all $x \in A$, x is $<$-minimal in A, or there is x' such that x' is $<$-minimal and $x' < x$.

Suppose our axiomatic theory T satisfies the following condition: if T is not classically consistent, then any inconsistency of T arises from a finite number of instances of its axioms (or axiom schemes). That is, if we retract a finite number of instances of the axioms (or axiom schemes) of an axiomatic system, the remaining instances form a consistent set. It is easy to verify that the prominent inconsistent theories in science and mathematics satisfy this condition. If it is not satisfied, a modular approach to paraconsistent reasoning loses its rationale. Suppose, furthermore, the set of all applications of axioms of T is countable. On these two assumptions, $<_s$ can be shown to be smooth in a straightforward manner[10]:

Proposition 1 *Suppose T satisfies the condition that, if T is not classically consistent, then any inconsistency of T arises from a finite number of applications of its axioms. Furthermore, the set of all applications of axioms of T is countable. Then, the satisfaction ordering $<_s$ of T is smooth in the set of $L(V)$ interpretations.*

See [1] for a proof of this proposition.

We can now set forth the inference relation of our semantics:

Definition 6 ($T \mid\!\sim \varphi$) Let T be a set of axioms upon which $<_s$ is defined by Definition 3. ϕ is an $L(V)$ formula. $T \mid\!\sim \varphi$ iff for all $<_s$-minimal worlds $\mathbf{w}$, $\mathbf{w} \models \varphi$.

Thus, φ is inferable from T iff φ is satisfied by all preferred global worlds of $L(V)$. This inference relation is intended to explicate reasoning from possibly inconsistent axiomatic theories.[11]

[10]The question of whether $<_s$ is smooth (in the set of all $L(V)$ interpretations) for axiomatic theories T with an uncountable set of applications must be left for future research.

[11]We should note that the presently defined inference relation deviates from the one defined in [10, 14]. They define the relation $A \mid\!\sim \varphi$ in such a way that φ must be verified by all classical models of A that are preferred on a given ordering $<$. This inference relation, however, is obviously not paraconsistent.

2.7 Prioritized Axiomatic Theories

The present semantics implies that if an application of an axiom α is not consistent with an application of an axiom β, neither application is accepted. This does not always accord with scientific practice. There are cases where some axiom is given priority to another axiom in the case of a conflict. For example, Planck's hypothesis of energy quantization is prioritized over certain applications of classical electrodynamics within Bohr's theory of the atom [1]. Such prioritizations are aptly characterized by a modular ordering among the axioms of T.[12] Such an ordering, in turn, can be represented by a sequence of sets of axioms (cf. [5]):

$$\Theta = \langle T_1, \ldots, T_m \rangle,$$

with the understanding that $\alpha \in T_p$ has priority over $\beta \in T_q$ iff $p < q$, where we assume that $\langle T_1, \ldots, T_m \rangle$ is a partition. All sets T_p $(1 \leq p \leq m)$ are sets of axioms. The union of these sets is the set of all axioms of the network under consideration.

We can take priorities among the axioms into account now by refining the satisfaction ordering among global worlds: a global world $\mathbf{w}$ is preferred over another global world $\mathbf{w}'$ iff, for some level p, $\mathbf{w}$ satisfies the axioms of T_p to a higher extent than $\mathbf{w}'$, without satisfying the axioms of the priority levels below p to a lesser extent. We first define the set of applications of axioms at level p that agree with the global world $\mathbf{w}$:

$$App_p(\mathbf{w}) =_{df} App(\mathbf{w}) \cap \{\mathcal{A}_{i,j} \mid \alpha_i \in T_p\}. \tag{5}$$

Now we can move on to defining the satisfaction ordering:

Definition 7 (*Satisfaction ordering* $<_s$ of Θ) Let $\mathbf{w}$ and $\mathbf{w}'$ be interpretations of the global language $L(V)$, and Θ a prioritized axiomatic theory. $\mathbf{w} <_s \mathbf{w}'$ iff

(1) there is some p $(1 \leq p \leq m)$ such that (i) $App_p(\mathbf{w}') \subset App_p(\mathbf{w})$, and
(2) for all $h < p$ $(h \geq 1)$, $App_h(\mathbf{w}') = App_h(\mathbf{w})$.

The satisfaction ordering $<_s$ is thus well defined for prioritized axiomatic theories Θ. Hence, Definition 6 can be adopted for such theories in a straightforward manner:

Definition 8 ($\Theta \mid\!\sim_< \varphi$) Let Θ be a set of prioritized axioms upon which $<_s$ is defined by Definition 7. φ is an $L(V)$ formula. $\Theta \mid\!\sim_< \varphi$ iff for all $<_s$-minimal worlds $\mathbf{w}$, $\mathbf{w} \models \varphi$.

---

[12] A strict partial order $<$ is modular iff the relation $R(x, y)$ defined by $x \not< y \wedge y \not< x$ is an equivalence relation. There might be cases where the axioms are not ordered in a modular fashion, which would require some modifications of Definition 7. For simplicity, a modular ordering among the axioms is assumed.

2.8 A Simple Example

It is time to exemplify our paraconsistent inference system. We shall give a highly simplified version of the inconsistency between Bohr's postulates and Maxwell's equations. It is a mere toy example but sufficiently rich to represent the core idea of the present formalism. When observing an inconsistency between Maxwell's equations and Bohr's postulates, we are concerned with the following general propositions (which are derivable from either Bohr's postulates or Maxwell's equation, together with some further pieces of background theory):

(1) If an electron is accelerated, it radiates.
(2) If an electron radiates, it loses energy.
(3) If an electron orbits around a proton, it does so at stable energy levels.[13]
(4) If an electron orbits around a proton, it is accelerated.
(5) If an electron circuits in an electromagnetic coil, it is accelerated.

Let us put these general propositions into formal axioms and introduce a formal language with the following symbols:

- $E(x)$: x is an electron.
- $O(x)$: x orbits around a proton.
- $W(x)$: x circuits in a wire of an electromagnetic coil.
- $C(x)$: x is accelerated.
- $R(x)$: x radiates.
- $L(x)$: x loses energy.

Our general claims can now be cast into a formal system T:

$$\forall x(E(x) \wedge C(x) \rightarrow R(x)), \qquad (\alpha_1)$$

$$\forall x(E(x) \wedge R(x) \rightarrow L(x)), \qquad (\alpha_2)$$

$$\forall x(E(x) \wedge O(x) \rightarrow \neg L(x)), \qquad (\alpha_3)$$

$$\forall x(E(x) \wedge O(x) \rightarrow C(x)), \text{ and} \qquad (\alpha_4)$$

$$\forall x(E(x) \wedge W(x) \rightarrow C(x)). \qquad (\alpha_5)$$

These axioms themselves are not inconsistent. However, if we add the claim that there is an electron that orbits around a proton, we obtain an inconsistency. Now, suppose we have two electrons e_1 and e_2. e_1 orbits around a proton, while e_2 circuits in a wire of an electromagnetic coil. These two electrons form applications of the axioms of our theory T. We can represent such applications by partial structures of the following types:

[13]This is a simplification because the energy levels are only relatively stable. An electron may jump from one level to another.

$$\langle A, E, C, R\rangle,$$

$$\langle A, E, R, L\rangle,$$

$$\langle A, E, O, L\rangle,$$

$$\langle A, E, O, C\rangle, \text{ and}$$

$$\langle A, E, W, C\rangle.$$

The domain A is always a singleton, containing either e_1 or e_2 in the present case. Our knowledge prior to applying the axioms merely consists in knowing that e_1 is an electron orbiting around a proton and e_2 an electron circuiting in an electromagnetic coil. The representation of this knowledge by partial structures of the above types is straightforward. For example, the partial structure representing the application of α_3 to e_1 is as follows:

$$\mathcal{A}_{3,1} = \langle \{e_1\}, \{e_1\}, \{e_1\}, \emptyset\rangle.$$

This partial structure has exactly one local world:

$$W(\mathcal{A}_{3,1}) =_{df} Mod(\alpha_3, (\mathcal{A}_{3,1})_1) \cap Ext(\mathcal{A}_{3,1}) = \{\langle \{e_1\}, \{e_1\}, \{e_1\}, \emptyset\rangle\}.$$

Which are the preferred global worlds that satisfy the axioms to a maximal extent? If we have no prioritization among the axioms, then there are four such worlds of the type $\langle A, E, O, W, C, R, L\rangle$:

$$\langle \{e_1, e_2\}, \{e_1, e_2\}, \{e_1\}, \{e_2\}, \{e_1, e_2\}, \{e_1, e_2\}, \{e_1, e_2\}\rangle,$$

$$\langle \{e_1, e_2\}, \{e_1, e_2\}, \{e_1\}, \{e_2\}, \{e_1, e_2\}, \{e_2\}, \{e_2\}\rangle,$$

$$\langle \{e_1, e_2\}, \{e_1, e_2\}, \{e_1\}, \{e_2\}, \{e_2\}, \{e_2\}, \{e_2\}\rangle, \text{ and}$$

$$\langle \{e_1, e_2\}, \{e_1, e_2\}, \{e_1\}, \{e_2\}, \{e_1, e_2\}, \{e_1, e_2\}, \{e_2\}\rangle.$$

Hence, without prioritization, we cannot derive any claim as to whether e_1 radiates or not. However, if we prioritize $\alpha_2, \alpha_3, \alpha_4$, and α_5 over α_1, only the second interpretation of $L(V)$ is a preferred global world. Hence, on the assumption of this prioritization, we obtain:

$$T \mathrel{|\!\sim} \neg R(e_1) \text{ and } T \mathrel{|\!\sim} R(e_2).$$

That is, the electron orbiting around the proton does not radiate, whereas the electron circuiting in the electromagnetic coil does radiate, as it should be.

Finally, let us explain why the original consolidation method by Rescher and Manor [13] does not suffice to account for our use of Bohr's postulates in the context of Maxwell's equations. Various inference relations can be defined on the basis of this consolidation method. Let $MCS(S)$ denote the set of maximal consistent subsets of S and $Free(S)$ denote the intersection of all maximal consistent subsets of S. Then, we can define (cf. [4]):

(1) $S \vdash_{free} \varphi$ iff $Free(S) \vdash_{CL} \varphi$,
(2) $S \vdash_{strong} \varphi$ iff for all $A \in MCS(S)$, $A \vdash_{CL} \varphi$,
(3) $S \vdash_{weak} \varphi$ iff there is $A \in MCS(S)$ s.t. $A \vdash_{CL} \varphi$, and
(4) $S \vdash_{argued} \varphi$ iff there is $A \in MCS(S)$ s.t. $A \vdash_{CL} \varphi$ but no $A \in MCS(S)$ s.t. $A \vdash_{CL} \neg\varphi$.

It is easy to verify then that none of the above inference relations yields the intended result. Suppose we have a non-prioritized premise set, containing Bohr's postulates, Maxwell's equations, and some further pieces of background theory. On the basis of $\vdash_{free}$ and $\vdash_{strong}$, no assertion can be derived as to whether e_1 radiates. Using $\vdash_{weak}$, we have to infer that e_1 does and does not radiate. With $\vdash_{argued}$ no assertion can be made as to whether e_1 radiates.

2.9 Modular Semantics Inference in Pure First Order Terms

Let $\mathcal{W}$ denote the set of formulas of first order classical logic and let $\mathcal{S} \subset \mathcal{W}$ denote literals (i.e. primitive formulas and their negations) of the same language. For each set A, we denote its cardinality by $|A|$.

In order to facilitate the translation of the Modular Semantics inference relations $\mid\!\sim$ and $\mid\!\sim_<$ into adaptive logics, we represent the above defined formal structures in a simple first order language. From now on (without loss of generality) we represent global worlds by Classical Logic models and partial structures $\mathcal{A}_{i,j}$ by combinations of a closed instance of a universally quantified formula (an axiom) and a corresponding finite set of literals (the known properties of the objects in the domain of the corresponding partial structure). For the latter we define a function $F(\alpha)$ that maps every instance of an axiom of a theory to a set of literals.

Let a *universal theory* be a set of formulas starting with the symbol $\forall$.

We now define the set of instantiations of a universal theory. $insts(\varphi) =_{df} \{\pi \mid \pi$ is an instance of $\varphi\}$ and, where T is a universal theory, $insts(T) = \bigcup\{insts(\varphi) \mid \varphi \in T\}$

The combination of a flat theory and its partial structures (henceforth called a *flat partially interpreted theory*) will thus be represented by a pair $\langle T, F\rangle$, where $T \subseteq \mathcal{W}$ and $F : insts(T) \to \mathcal{P}(\mathcal{S})$. Let FIT be the set of all flat partially interpreted theories, i.e. all such pairs. The inference relation $\mid\!\sim$ will be represented by a syntactic relation $\vdash_{FN}$ in $FIT \times \mathcal{W}$.

With the following three definitions we give the definition of $\vdash_{FN}$.

Definition 9 $App(M, \langle T, F\rangle) =_{df} \{\varphi \mid \varphi \in insts(T)$ and $M \models \{\varphi\} \cup F(\varphi)\}$.

Definition 10 $M <_{\langle T,F\rangle} M' =_{df} App(M', \langle T, F\rangle) \subset App(M, \langle T, F\rangle)$.

Definition 11 $\langle T, F\rangle \vdash_{FN} \varphi =_{df} M \models \varphi$, for all $<_{\langle T,F\rangle}$-minimal elements M in all models.

A little thought shows that in fact this inference relation can also be expressed in terms of maximal consistent subsets.

Theorem 1 $\langle T, F\rangle \vdash_{FN} \varphi$ *iff* $\{\psi \wedge \bigwedge F(\psi) \mid \psi \in insts(T)\} \vdash_{strong} \varphi$.

Let us now proceed to the prioritized version of our inference relations. The combination of a prioritized theory and its partial structures (henceforth called a *prioritized partially interpreted theory*) will be represented by a pair $\langle\langle T_1, \ldots, T_n\rangle, F\rangle$, where each $T_i \subseteq \mathcal{W}$ and $F : insts(T_1 \cup \ldots \cup T_n) \to \mathcal{P}(\mathcal{S})$. Let PIT be the set of all prioritized partially interpreted theories, i.e. all such pairs. The inference relation $\mid\!\sim_{<}$ will be represented by a syntactic relation $\vdash_{PN}$ in $PIT \times \mathcal{W}$.

With the following three definitions we give the definition of $\vdash_{PN}$.

Definition 12 Where $i \leq m$, $App_i(M, \langle\langle T_1, \ldots, T_m\rangle, F\rangle) =_{df} \{\varphi \mid \varphi \in insts(T_i)$ and $M \models \{\varphi\} \cup F(\varphi)\}$.

Definition 13 Where $\Theta = \langle T_1, \ldots, T_m\rangle$, $M <_{\langle\Theta,F\rangle} M' =_{df}$ there is a $p \leq m$ such that $App_p(M', \langle\Theta, F\rangle) \subset App_p(M, \langle\Theta, F\rangle)$ and, for all $h < p$, $App_h(M, \langle\Theta, F\rangle) = App_h(M', \langle\Theta, F\rangle)$.

Definition 14 Where $\Theta = \langle T_1, \ldots, T_m\rangle$, $\langle\Theta, F\rangle \vdash_{PN} \varphi =_{df} M \models \varphi$, for all $<_{\langle\Theta,F\rangle}$-minimal elements M in all models.

2.10 The Example Revisited: Now in First Order Terms

In the example from Sect. 2.8, $insts(T)$ is the set of instances of α_1–α_5 once with the constant e_1 and once with e_2. We write the instance of α_i with e_j as $\psi_{i,j}$, e.g. $\psi_{3,1} = (E(e_1) \wedge O(e_1)) \to \neg L(e_1)$ and $\psi_{4,2} = (E(e_2) \wedge O(e_2)) \to C(e_2)$. With this notation we can write the instances of T as follows:

$$insts(T) = \{\psi_{i,j} \mid 1 \leq i \leq 5, j \in \{1, 2\}\}.$$

Given the information provided about e_1 and e_2, we moreover have

$$F(\psi_{i,j}) = \begin{cases} \{E(e_1), O(e_1), \neg W(e_1)\} & \text{if } j = 1 \\ \{E(e_2), \neg O(e_2), W(e_2)\} & \text{if } j = 2. \end{cases}$$

We use the following shorthand:

$$\varphi_{i,j} =_{df} \psi_{i,j} \wedge \bigwedge F(\psi_{i,j}).$$

This notation convention gives us e.g.

$$\varphi_{4,2} = ((E(e_2) \wedge O(e_2)) \rightarrow C(e_2)) \wedge E(e_2) \wedge \neg O(e_2) \wedge W(e_2).$$

We obtain the $\vdash_{FN}$-consequences of $\langle T, F\rangle$ by selecting the maximal consistent subsets of $\{\varphi_{i,j} \mid 1 \leq i \leq 5, j \in \{1, 2\}\}$. There is no inconsistency for e_2 or for the instance of α_5 with e_1, so each of the maximal consistent subsets contain all members of $\{\varphi_{i,2} \mid 1 \leq i \leq 5\} \cup \{\varphi_{5,1}\}$. The other members of the four maximal consistent subsets are respectively

$$\{\varphi_{1,1}, \varphi_{2,1}, \varphi_{4,1}\},$$
$$\{\varphi_{2,1}, \varphi_{3,1}, \varphi_{4,1}\},$$
$$\{\varphi_{1,1}, \varphi_{2,1}, \varphi_{3,1}\}, \text{ and}$$
$$\{\varphi_{1,1}, \varphi_{3,1}, \varphi_{4,1}\}.$$

Each of these maximal consistent subsets have the corresponding (given the same order of mentioning) preferred world on page 13 as a model. From all maximal consistent subsets the following literals are derivable (in classical logic)

$$B = \{E(e_1), O(e_1), \neg W(e_1), E(e_2), \neg O(e_2), W(e_2), C(e_2), R(e_2), L(e_2)\}. \quad (6)$$

but they do not agree on the other literals. The following literals are respectively derivable from the maximal consistent subsets:

$$\{L(e_1), R(e_1), C(e_1)\},$$
$$\{\neg L(e_1), \neg R(e_1), C(e_1)\},$$
$$\{\neg L(e_1), \neg R(e_1), \neg C(e_1)\}, \text{ and}$$
$$\{\neg L(e_1), R(e_1), C(e_1)\}.$$

The $\vdash_{FN}$-consequences are now the classical logic consequences of the intersection of all maximal consistent subsets, among which are the classical logic consequences of B as defined in expression (6). We can again observe that the non-prioritized version of the inference relation does not give us information on $R(e_1)$. Neither $R(e_1)$ nor its negation are derivable from $\langle T, F\rangle$.

However, the prioritization suggested in Sect. 2.8, allows us to derive more. We now calculate the $\vdash_{PN}$-consequences of $\langle\langle\{\alpha_2, \alpha_3, \alpha_4, \alpha_5\}, \{\alpha_1\}\rangle, F\rangle$. This prioritization allows for a more narrow selection of maximal consistent subsets. In fact only the second subset, i.e. the one that is a superset of

$$\{\varphi_{2,1}, \varphi_{3,1}, \varphi_{4,1}\}$$

does not verify the less important α_1 at the price of falsifying the more important α_2, α_3 or α_4.

We can therefore conclude that

$$\langle\langle\{\alpha_2, \alpha_3, \alpha_4, \alpha_5\}, \{\alpha_1\}\rangle, F\rangle \vdash_{PN} \varphi$$

iff

$$\{\varphi_{2,1}, \varphi_{3,1}, \varphi_{4,1}, \varphi_{5,1}, \varphi_{1,2}, \varphi_{2,2}, \varphi_{3,2}, \varphi_{4,2}, \varphi_{5,2}\} \vdash \varphi.$$

This makes $R(e_2)$ and $\neg R(e_1)$ paraconsistent consequences of our prioritized inconsistent theory.

3 Adaptive Proofs

Up to this point we have obtained a nice characterization of two paraconsistent inference relations, one flat and one prioritized. We will now define a dynamic proof theory that characterizes these relations. This proof theory is the proof theory of adaptive logics. We first give a short introduction to adaptive logics, then we characterize the inference relations by two semantically characterized adaptive logics, and finally we give the proof theory of these adaptive logics.

3.1 Adaptive Logics: An Introduction

Adaptive logics are logics which provide one unified proof theory for a wide range of non-monotonic consequence relations (cf. [3, 16]). Adaptive logics were devised by Diderik Batens to formalize defeasible reasoning forms. Each adaptive logic is conceived as a consequence relation in between two Tarski consequence relations, a *lower limit logic* and a stronger *upper limit logic*. Depending on the set of premises, an adaptive logic adds to all the instances of the lower limit logic rules as many of the instances of the rules of the upper limit logic as possible. This is done in such a way that each premise set has as many consequences as possible without being trivialized.

The upper limit logic is in fact defined as the lower limit logic plus an axiom schema stating that a specific set of lower limit logic contingent[14] formulas of a specific form are false. This specific set is called the set of *abnormalities*. So adaptive logics can be characterized as the lower limit logic plus the assumption that as many abnormalities as possible are false, given the restrictions imposed by the premises.

The proofs of adaptive logics are dynamic in nature: the lines of the proofs are conditional and their derivation status may change from not derived to derived and back to not derived as the proof goes on. This dynamics formalizes the increasing

[14] Where **L** is a logic, a formula φ is **L**-*contingent* iff there is an **L**-model that falsifies φ and one that verifies φ.

insight in the premises and how this may force one to conclude that former conditional derivations are no longer acceptable. The actual adaptive logic consequences, i.e. intuitively a set of derivable consequences which can be defended against every criticism, are exactly specified for every premise set and therefore the adaptive consequence relation is always a well defined relation. One might therefore say that adaptive proofs converge to the derivation of the correct adaptive consequences.

This syntactically defined adaptive consequence relation is equivalent to the semantic adaptive consequence relation. The latter is defined by selecting those lower limit logic models of the premises that verify the least number of abnormalities. The consequences are the formulas that are verified by all selected models of the premises.

There are several generic formats (with different degrees of generality) for adaptive logics. This means that in order to create a specific adaptive logic, one only needs to define the logic by a few parameters. The definition of its semantics and proof theory and the proofs of the most important metatheorems come for free with the generic formats. The first and least general format is called the *standard format of adaptive logics* (cf. [3]). Later on the more general *format of lexicographic adaptive logics* was developed (cf. [12]). This was further generalized to the *threshold functions format of adaptive logics* (cf. [16]).

The first logic we will define will be in standard format and both are in lexicographic format (and therefore also in threshold functions format). We start with a semantic characterization of a relevant subset of the adaptive logics in standard and in lexicographic format (those with (lexicographic) Minimal Abnormality strategy).

3.2 The Semantics of the Relevant Classes of Adaptive Logics

We first give a semantic definition of the minimal abnormality adaptive logics in standard format. We define them here as a pair. Note that one normally defines adaptive logics in standard format as a triple, because in the full format one can also vary the selection strategy of the logic. This further variation is of no use for the purpose of this paper. In what follows, let $\mathbb{N}$ denote the set of natural numbers.

Definition 15 A *minimal abnormality AL* is a pair consisting of

(1) a *lower limit logic* **LLL**: a reflexive, transitive, monotonic and compact consequence relation.
(2) a set of *abnormalities* A: a set of **LLL**-contingent formulas, characterized by a logical form.

Where M is an **LLL**-model, $Ab(M) =_{df} \{\varphi \in A \mid M \models \varphi\}$.

Definition 16 'Less abnormal' ordering. $M <_m M' =_{df} Ab(M) \subset Ab(M')$.

Definition 17 Least abnormal models. $\mathcal{M}^m(T) =_{df}$ the elements that are $<_m$-minimal in $\{M$ is an **LLL**-model $\mid M \models T\}$.

Definition 18 $T \vDash_{\mathbf{AL}^m} \varphi$ iff, for all $M \in \mathcal{M}^m(T)$, $M \models \varphi$.

We now semantically define the minimal abnormality lexicographic logics in lexicographic format. Also here we need to remark that we reduce the usual triple to a pair because we do not need to consider other strategies than lexicographic minimal abnormality.

Definition 19 A *lexicographic minimal abnormality AL* is a pair consisting of

(1) a *lower limit logic* **LLL**: a reflexive, transitive, monotonic and compact consequence relation.
(2) a sequence of *abnormalities* $\langle A_i \rangle_{i \in I}$, where I is $\mathbb{N}$ or an initial fragment of $\mathbb{N}$; each A_i is a set of **LLL**-formulas, characterized by a logical form.

Where $i \in I$ and M a **LLL**-model, $Ab_i(M) =_{df} \{\varphi \in A_i \mid M \models \varphi\}$.

Definition 20 Lexicographic 'less abnormal' ordering. Where M and M' are **LLL**-models, $M \sqsubset_l M' =_{df}$ there is a $j \in I$ such that $Ab_j(M) \subset Ab_j(M')$ and, for all $i < j$, $Ab_i(M) = Ab_i(M')$.

Definition 21 Lexicographically least abnormal models. $\mathcal{M}^l(T) =_{df}$ the elements that are $\sqsubset_l$-minimal in $\{M$ is an **LLL**-model $\mid M \models T\}$.

Definition 22 $T \vDash_{\mathbf{AL}^l} \varphi$ iff, for all $M \in \mathcal{M}^l(T)$, $M \models \varphi$.

3.3 Adaptive Logics for the Modular Semantics Inference Relations

We will now define two adaptive logics that capture the inference relations $\vdash_{PN}$ and $\vdash_{FN}$ in a similar way as Rescher Manor consequence is characterized by means of an adaptive logic in [11, 16]. We could in principle have used Theorem 1 and the existing translation of the Strong consequences to adaptive logic to obtain adaptive proofs for the inference relations, but it seems useful to do the translation from the Modular Semantics inference relations to adaptive logic in a direct way. Moreover it is useful to observe the parallels between the flat and the prioritized case in the metaproofs we provide below.[15]

Before we are ready to define the adaptive logics, we need to define the logic $\mathbf{CL}^\circ$, which will function as the **LLL** of the adaptive logics we will present. $\mathbf{CL}^\circ$ is basically classical first order logic with a set of dummy operators $\circ$ and $\circ^i$ where $i \in \mathbb{N}$. One could interpret these dummy operators as symbols that reduce the formula which they precede to a black box, i.e. a solid proposition that cannot be analyzed

[15] We hope it is clear for the reader that we do not claim to define a new adaptive logic here, but rather apply existing logics to provide a new solution for an interesting problem, viz. providing a proof theory for the Modular Semantics inference relations.

any further and is true or false independent of its structure. The language $\mathcal{L}$ of **CL**$^\circ$ is therefore simply the language of classical logic with additional unary connectives $\circ$ and $\circ^i$. The set of closed formulas of $\mathcal{L}$ will be denoted by $\mathcal{W}$. $\mathcal{C}$ and $\mathcal{P}_r$ respectively denote the constants and the r-ary predicates of $\mathcal{L}$.

For each cardinality $\mathfrak{c}$ we define a pseudo-language $\mathcal{L}^\mathfrak{c}$. Let $\mathcal{O}^\mathfrak{c}$ be a set of pseudo-constants of cardinality $\mathfrak{c}$. $\mathcal{L}^\mathfrak{c}$ is the result of extending $\mathcal{L}$ by replacing $\mathcal{C}$ by $\mathcal{C} \cup \mathcal{O}^\mathfrak{c}$ in the construction of the formulas $\mathcal{W}^\mathfrak{c}$ of $\mathcal{L}^\mathfrak{c}$.

A **CL**$^\circ$-model M is a pair $\langle D, v\rangle$, where v is the union of the following functions:

$v' : \mathcal{W}^{|D|} \to \{0, 1\}$,
$v'' : \mathcal{C} \cup \mathcal{O}^{|D|} \to D$, and
$v''' : \mathcal{P}_r \to D^r$.

Every model $M = \langle D, v\rangle$ uniquely defines a valuation function $v_M : \mathcal{W}^{|D|} \to \{0, 1\}$, recursively defined by the following clauses:

$v_M(P^r\alpha_1 \ldots \alpha_r) = 1$ iff $\langle v(\alpha_1) \ldots v(\alpha_r)\rangle \in v(P^r)$,
$v_M(\forall\alpha\varphi(\alpha)) = 1$ iff $v_M(\varphi(\beta)) = 1$ for all $\beta \in \mathcal{C} \cup \mathcal{O}^{|D|}$,
$v_M(\neg\varphi) = 1$ iff $v_M(\varphi) = 0$,
$v_M(\varphi \vee \pi) = 1$ iff $v_M(\varphi) = 1$ or $v_M(\pi) = 1$,
$v_M(\circ A) = 1$ iff $v(\circ A) = 1$, and
$v_M(\circ^i A) = 1$ iff $v(\circ^i A) = 1$.

The symbols $\exists$, $\wedge$, and $\to$ are defined from the other symbols in the usual way.

A model M satisfies a formula φ, written $M \models \varphi$, iff $v_M(\varphi) = 1$. From this the semantic consequence relation $\models$ of **CL**$^\circ$ is defined as usual. A proof theory can be obtained by extending the axioms and rules of an axiomatization of Classical Logic to $\mathcal{W}$. No rules or axioms for $\circ$ and $\circ^i$ have to be added. This can be seen by considering that $\circ\varphi$ is supposed to be a black box, the contents of which cannot be used in derivations. One can do exactly the same with $\circ\varphi$ as with a primitive formula. Given that there are no rules for primitive formulas, one does not need rules for $\circ$ either.

Now we have all ingredients needed to define the adaptive logics for $\mid\!\sim$ and $\mid\!\sim_<$.

Definition 23 The adaptive logic **ALF** is the Minimal Abnormality adaptive logic defined by **LLL** $=$ **CL**$^\circ$ and set of abnormalities $A = \{\circ\varphi \wedge \neg\varphi \mid \varphi \in \mathcal{W}\}$.

Definition 24 The adaptive logic **ALP** is the Lexicographic Minimal Abnormality adaptive logic defined by **LLL** $=$ **CL**$^\circ$ and sequence of abnormalities $\langle A_i\rangle_{i\in\mathbb{N}}$ where $A_i = \{\circ^i\varphi \wedge \neg\varphi \mid \varphi \in \mathcal{W}\}$ for all $i \in \mathbb{N}$.

In order to apply the adaptive logics correctly one needs a translation of the theories into a form with $\circ$ and $\circ^i$ symbols. For this purpose we need the following definitions.

Before we get to the first crucial adequacy theorem, we need to prove the following lemma.

Lemma 1 *If (a)* $M \models \circ\varphi$ *iff* $\varphi = \psi \wedge \bigwedge F(\psi)$ *for some* $\psi \in insts(T)$ *and (b)* $M' \models \circ\varphi$ *iff* $\varphi = \psi \wedge \bigwedge F(\psi)$ *for some* $\psi \in insts(T)$*, then*

$$M <_{\langle T,F\rangle} M' \text{ iff } M <_m M',$$

where $<_m$ *is the less abnormal order relation of the adaptive logic* **ALF***.*

Proof Suppose the antecedent of the lemma holds.

$$M <_{\langle T,F\rangle} M'$$

$\Leftrightarrow$

$$\{\varphi \mid \varphi \in insts(T) \text{ and } M' \models \{\varphi\} \cup F(\varphi)\} \subset \{\varphi \mid \varphi \in insts(T) \text{ and } M \models \{\varphi\} \cup F(\varphi)\}$$

$\Leftrightarrow$

$$\{\varphi \mid \varphi \in insts(T) \text{ and } M \not\models \{\varphi\} \cup F(\varphi)\} \subset \{\varphi \mid \varphi \in insts(T) \text{ and } M' \not\models \{\varphi\} \cup F(\varphi)\}$$

$\Leftrightarrow$ (in view of $\varphi \in insts(T)$ iff $M \models \circ(\varphi \wedge \bigwedge F(\varphi))$, whenever $M \models \{\circ(\psi \wedge \bigwedge F(\psi)) \mid \psi \in insts(T)\}$)

$$\{\varphi \mid M \models \circ(\varphi \wedge \bigwedge F(\varphi)) \text{ and } M \models \neg(\varphi \wedge \bigwedge F(\varphi))\} \subset \\ \{\varphi \mid M' \models \circ(\varphi \wedge \bigwedge F(\varphi)) \text{ and } M' \models \neg(\varphi \wedge \bigwedge F(\varphi))\}$$

$\Leftrightarrow$ (because, ψ is of the form $\varphi \wedge \bigwedge F(\varphi)$ whenever $M \models \circ\psi$ or $M' \models \circ\psi$)

$$\{\varphi \mid M \models \circ\varphi \wedge \neg\varphi\} \subset \{\varphi \mid M' \models \circ\varphi \wedge \neg\varphi\}$$

$\Leftrightarrow$

$$Ab(M) \subset Ab(M')$$

$\Leftrightarrow$

$$M <_m M'.$$

□

Now we have all the means to prove the adequacy theorem for the flat case.

Theorem 2 *Where* T*,* φ*, and, for all* ψ*,* $F(\psi)$ *do not contain* $\circ$ *or* $\circ^i$*,* $\{\circ(\psi \wedge \bigwedge F(\psi)) \mid \psi \in insts(T)\} \models_{ALF} \varphi$ *iff* $\langle T, F\rangle \vdash_{FN} \varphi$*.*

Proof Suppose the antecedent is true.

In this proof, the variables M, M', M'' and M''' will always refer to **CL**$^\circ$-models. Let G be the set $\{\circ(\psi \wedge \bigwedge F(\psi)) \mid \psi \in insts(T)\}$. Let M be G-*canonical* (abbreviated to $Ca(M)$) iff $M \models \circ\psi$ iff $\circ\psi \in G$.

First observe that, for every M, there is a G-canonical M' such that M' verifies exactly the same $\circ$-free formulas as M (Observation 1). This is because we can construct a G-canonical model from M by only changing the truth values of the formulas of the form $\circ\psi$ (make those in G true and the other ones false). This is possible because these formulas always receive independent truth values.

Hence, where ψ is $\circ$-free we have (where $\mathcal{M}_1$ is any set of $\mathbf{CL}^\circ$-models):

$M \models \psi$, for all $M \in \mathcal{M}_1$,
iff
$M \models \psi$, for all $\{M \in \mathcal{M}_1 \mid Ca(M)\}$ (call this Observation 2).

Next, observe that it is impossible that there is a non-G-categorical M such that $M \models G$ and a G-categorical M' such that $M <_m M'$, but no G-categorical M'' such that $M'' <_m M'$ (call this Observation 3). If it were possible, we would be able to transform M into a G-categorical model M'' such that the formulas of the form $\circ\psi \wedge \neg\psi$ it verifies would be a subset of those verified by M. This M'' is constructed as the model identical to M except that the truth values of formulas $\circ\psi \notin G$ are false. M and M'' verify the same $\circ$-free formulas of the form $\neg\psi$ and M'' verifies a proper subset of formulas of the form $\circ\psi$ compared to M. M'' therefore does not verify more formulas $\circ\psi \wedge \neg\psi$ than M. Hence $M'' <_m M'$, and M'' is G-categorical, which contradicts our hypothesis. Observation 3 thus holds, which implies that adding non-G-categorical models to a set of G-categorical models from which the minimal ones are picked, can never make a $<_m$-minimal model non-$<_m$-minimal.
Let $\min_<(A) =_{df}$ the set of $<$-minimal elements in A.
By Lemma 1 we know that

if $Ca(M_1)$ and $Ca(M_2)$, then ($M_1 <_m M_2$ iff $M_1 <_{\langle T,F\rangle} M_2$)
$\Rightarrow$
$\min_{<_m}(\{M \mid Ca(M)\}) = \min_{<_{\langle T,F\rangle}}(\{M \mid Ca(M)\})$
$\Rightarrow$
($M \models \varphi$, for all $M \in \min_{<_m}(\{M \mid Ca(M)\})$)
iff
($M \models \varphi$, for all $M \in \min_{<_{\langle T,F\rangle}}(\{M \mid Ca(M)\})$).
$\Rightarrow$ (in view of Observation 2 and 3)
($M \models \varphi$, for all $M \in \min_{<_m}(\{M \mid M \models G\})$)
iff
($M \models \varphi$, for all $M \in \min_{<_{\langle T,F\rangle}}(\{M \mid Ca(M)\})$).

$\Rightarrow$ (in view of the fact that the $<_{\langle T,F\rangle}$ ordering has nothing to do with $\circ$-formulas—one could say that they are innocent bystanders; the restriction to G-categorical models has no effect whatsoever on the ordering and the verified formulas)
($M \models \varphi$, for all $M \in \min_{<_m}(\{M \mid M \models G\})$)
iff
($M \models \varphi$, for all $M \in \min_{<_{\langle T,F\rangle}}(\{M \mid M$ is a **CL**-model$\})$).
$\Rightarrow$
$G \vDash_{ALF} \varphi$ iff $\langle T, F\rangle \vdash_{FN} \varphi$. □

For the adequacy of the prioritized case we also need to first prove a lemma.

Lemma 2 *If (a)* $M \models \circ^i\varphi$ *iff* $\varphi = \psi \wedge \bigwedge F(\psi)$ *for some* $\psi \in insts(T_i)$ *and (b)* $M' \models \circ^i\varphi$ *iff* $\varphi = \psi \wedge \bigwedge F(\psi)$ *for some* $\psi \in insts(T_i)$*, then*

$$M <_{\langle\Theta,F\rangle} M' \text{ iff } M \sqsubset_l M',$$

where $\sqsubset_l$ *is the lexicographic less abnormal order relation of the adaptive logic* **ALP**.

Proof Suppose the antecedent of the lemma holds.

$$M <_{\langle\Theta,F\rangle} M'$$

$\Leftrightarrow$

there is a $p \leq m$ such that

$$\{\varphi \mid \varphi \in insts(T_p) \text{ and } M' \models \{\varphi\} \cup F(\varphi)\} \subset \{\varphi \mid \varphi \in insts(T_p) \text{ and } M \models \{\varphi\} \cup F(\varphi)\}$$

and, for all $h < p$,

$$\{\varphi \mid \varphi \in insts(T_h) \text{ and } M' \models \{\varphi\} \cup F(\varphi)\} = \{\varphi \mid \varphi \in insts(T_h) \text{ and } M \models \{\varphi\} \cup F(\varphi)\}$$

$\Leftrightarrow$

there is a $p \leq m$ such that

$$\{\varphi \mid \varphi \in insts(T_p) \text{ and } M \not\models \{\varphi\} \cup F(\varphi)\} \subset \{\varphi \mid \varphi \in insts(T_p) \text{ and } M' \not\models \{\varphi\} \cup F(\varphi)\}$$

and, for all $h < p$,

$$\{\varphi \mid \varphi \in insts(T_h) \text{ and } M \not\models \{\varphi\} \cup F(\varphi)\} = \{\varphi \mid \varphi \in insts(T_h) \text{ and } M' \not\models \{\varphi\} \cup F(\varphi)\}$$

$\Leftrightarrow$ (in view of $\varphi \in insts(T_n)$ iff $M \models \circ^n(\varphi \wedge \bigwedge F(\varphi))$, whenever $M \models \{\circ^i(\psi \wedge \bigwedge F(\psi)) \mid \psi \in insts(T_i); i \leq n\}$)

there is a $p \leq m$ such that

$$\{\varphi \mid M \models \circ^p(\varphi \wedge \bigwedge F(\varphi)) \text{ and } M \models \neg(\varphi \wedge \bigwedge F(\varphi))\} \subset$$
$$\{\varphi \mid M' \models \circ^p(\varphi \wedge \bigwedge F(\varphi)) \text{ and } M' \models \neg(\varphi \wedge \bigwedge F(\varphi))\}$$

and, for all $h < p$,

$$\{\varphi \mid M \models \circ^h(\varphi \wedge \bigwedge F(\varphi)) \text{ and } M \models \neg(\varphi \wedge \bigwedge F(\varphi))\} =$$
$$\{\varphi \mid M' \models \circ^h(\varphi \wedge \bigwedge F(\varphi)) \text{ and } M' \models \neg(\varphi \wedge \bigwedge F(\varphi))\}$$

$\Leftrightarrow$ (because ψ is of the form $\varphi \wedge \bigwedge F(\varphi)$ whenever $M \models \circ^i\psi$ or $M' \models \circ^i\psi$), there is a $p \leq m$ such that

$$\{\varphi \mid M \models \circ^p\varphi \wedge \neg\varphi\} \subset \{\varphi \mid M' \models \circ^p\varphi \wedge \neg\varphi\}$$

and, for all $h < p$,

$$\{\varphi \mid M \models \circ^h\varphi \wedge \neg\varphi\} = \{\varphi \mid M' \models \circ^h\varphi \wedge \neg\varphi\}$$

$\Leftrightarrow$

there is a $p \leq m$ such that

$$Ab_p(M) \subset Ab_p(M')$$

and, for all $h < p$,

$$Ab_h(M) \subset Ab_h(M')$$

$\Leftrightarrow$

$$M \sqsubset_l M'.$$

□

We can now prove that the prioritized adaptive logic is adequate w.r.t. the prioritized inference relation.

Theorem 3 *Where* $\Theta = \langle T_1, \ldots, T_m\rangle$, φ, *and, for all* ψ, $F(\psi)$ *do not contain* $\circ$ *or* $\circ^i$, $\{\circ^i(\psi \wedge \bigwedge F(\psi)) \mid \psi \in insts(T_i); i \leq m\} \vDash_{ALP} \varphi$ *iff* $\langle\Theta, F\rangle \vdash_{PN} \varphi$.

Proof The result can be obtained from Lemma 2 in a very similar way as Theorem 2 was obtained from Lemma 1. □

At this point some readers may be disappointed that we do not define an adaptive logic that starts directly from premises that are solely based on the axioms of the inconsistent theories (e.g. with a subclassical lower limit logic (weakening the universal quantifier) and abnormalities $\forall x\pi(x) \wedge \neg\pi(a)$—we are indebted to a referee for this suggestion). This may be an interesting project for the future, but for now better results seem to be obtainable if we prepare our theory with some extralogical information (the information contained in the function F).

3.4 Adaptive Proofs: Definitions

We are now ready to present the dynamic proofs of the adaptive logics **ALF** and **ALP**. The former logic is an adaptive logic in standard format, so we can simply follow the standard proofs of this format. The latter is a standard lexicographic adaptive logic so we can also borrow the proofs of this format. The proofs of **ALF** and **ALP** have most of their structure in common, the only difference being that they have a different way of determining which *choice sets* are *minimal* and which are not. In what follows the metavariable A refers either to the set of abnormalities of **ALF** or to the set

$\bigcup\{A_i \mid i \in \mathbb{N}\}$, where $\langle A_i \rangle_{i \in \mathbb{N}}$ is the sequence of abnormalities of **ALP**, depending on the logic we are using. Where C is a finite set of abnormalities, $\text{Dab}(C) = \bigvee C$.

The lines of **AL**-proofs have four elements: a line number i, a formula φ, a justification (a derivation rule and the lines to which this rule is applied), and a condition $D \subseteq A$. Where T is the set of premises, the inference rules are defined in the following table (we omit the line numbers and justifications).

PREM If $\varphi \in T$:

$$\frac{\vdots \quad \vdots}{\varphi \quad \emptyset}$$

RU If $\varphi_1, \ldots, \varphi_n \vdash_{\mathbf{CL}^\circ} \psi$:

$$\frac{\begin{array}{ll}\varphi_1 & D_1 \\ \vdots & \vdots \\ \varphi_n & D_n\end{array}}{\psi \quad D_1 \cup \ldots \cup D_n}$$

RC If $\varphi_1, \ldots, \varphi_n \vdash_{\mathbf{CL}^\circ} \psi \vee \text{Dab}(C)$

$$\frac{\begin{array}{ll}\varphi_1 & D_1 \\ \vdots & \vdots \\ \varphi_n & D_n\end{array}}{\psi \quad D_1 \cup \ldots \cup D_n \cup C}$$

The premise rule (PREM) enables the introduction of premises; the unconditional rule (RU) enables the derivation of **CL**°-consequences from formulas on preceding lines; the conditional rule (RC) allows one to push abnormalities to the condition. So the rule RC allows us to derive φ from $\circ\varphi$ in a proof from T. One has thus obtained the derived rule RD.

RD $$\frac{\circ\varphi \quad D}{\varphi \quad D \cup \{\circ\varphi \wedge \neg\varphi\}}$$

The adaptive proofs are dynamic. This means that the derivation status (whether the formulas of the lines are considered as derived) may change as the proof goes on. We call a line *marked* if its formula is not considered derived and *unmarked* if its formula is considered derived. Lines can be marked and later unmarked, and yet later marked again. So with every added line the derivation status of the original lines may also change. For this reason we do not define an adaptive proof as a list of lines but as a chain of so-called *stages*, i.e. lists of lines. Adding a line to a proof by applying one of the above rules brings the proof to its next stage, which is the sequence of all lines written so far.

At every stage of a proof, a marking definition determines for each line in the proof whether it is marked or not. If a line with formula φ is marked at stage s, this indicates that given our best insights at this stage, φ cannot be considered derived on that line. If the line is unmarked at stage s, we say that φ is derived at stage s of the proof.

Where $D \subset A$, *Dab*(D) is a *Dab-formula* at stage s of a proof iff it is the second element of a line at stage s with an empty condition. A more direct way to derive Dab-formulas is by the application of the following derived rule.

MR If $\varphi_1, \ldots, \varphi_n \vdash_{\mathbf{CL}^\circ} \bot$:

$$\begin{array}{ll} \varphi_1 & D_1 \\ \vdots & \vdots \\ \varphi_n & D_n \\ \hline Dab(D_1 \cup \ldots \cup D_n) & \emptyset \end{array}$$

MR can be understood as follows. We have conditionally derived $\varphi_1, \ldots, \varphi_n$, but we find out later that these formulas are together inconsistent. Hence, there must be something wrong with at least one of the conditions on which we derived them. Hence we derive the Dab-formula $Dab(D_1 \cup \ldots \cup D_n)$.

$Dab(D)$ is a *minimal* Dab-formula at stage s iff there is no other Dab-formula $Dab(D')$ at stage s for which $D' \subset D$. Where $Dab(D_1), Dab(D_2), \ldots$ are the minimal Dab-formulas derived at stage s, let $\Sigma_s =_{df} \{D_1, D_2, \ldots\}$. We say that $C \subseteq A$ is a *choice set* of Σ_s iff C contains an element of every $D_i \in \Sigma_s$. Let $\sqsubset$ be the order relation on sets of abnormalities that is defined as follows: $C_1 \sqsubset C_2 =_{df}$ there is a $j \in I$ such that $(C_1 \cap A_j) \subset (C_2 \cap A_j)$ and, for all $i < j$, $(C_1 \cap A_i) = (C_2 \cap A_i)$.

Finally we arrive at the only difference between **ALF** and **ALP**: the way in which the set Φ_s of minimal choice sets of the minimal Dab-formulas at a stage of a proof is determined. Define Φ_s of stage s of an **ALF**-proof respectively a **ALP**-proof as the $\subset$-minimal resp. the $\sqsubset$-minimal elements in the set of all choice sets of Σ_s.

Now we can determine when a line is marked at a stage of a proof.

Definition 25 A line with formula φ and condition D is marked at stage s of an **AL**-proof from Γ iff (i) there is no $C \in \Phi_s$ such that $C \cap D = \emptyset$, or (ii) for some $C \in \Phi_s$, φ is not derived on a condition D at stage s such that $C \cap D = \emptyset$.

Intuitively a line is marked when, given the present insights in the premises, at least one of the presuppositions of that line (represented by the negation of the formulas in the condition element of the line) is unreasonable.

For example, suppose one line in a proof at a stage has as its condition element $\{\circ p \wedge \neg p, \circ q \wedge \neg q\}$ and another line's condition is $\{\circ r \wedge \neg r\}$. The lines have different formulas. Suppose moreover that $(\circ p \wedge \neg p) \vee (\circ r \wedge \neg r) \vee (\circ s \wedge \neg s)$ is a minimal Dab-formula at the stage. We now know that, given the truth of the premises, it is inconsistent to presuppose $\circ p \rightarrow p$, $\circ r \rightarrow r$, and $\circ s \rightarrow s$ together. Because we have no way to choose one of the three, they should all be considered problematic. One will therefore mark both conditional lines, as they each have a problematic presupposition.

The situation is different when the two lines have the same formula φ. Consider the minimally abnormal interpretations of $(\circ p \wedge \neg p) \vee (\circ r \wedge \neg r) \vee (\circ s \wedge \neg s)$. In each such interpretation $\circ q \rightarrow q$ is true and either (1) $\circ p \rightarrow p$ and $\circ s \rightarrow s$ are true, or (2) $\circ p \rightarrow p$ and $\circ r \rightarrow r$ are true, or (3) $\circ s \rightarrow s$ and $\circ r \rightarrow r$ are true. In case (1) the presuppositions of the first line are satisfied. In case (2) and (3) the presuppositions of the second line are satisfied. Hence, in each minimally abnormal situation, φ is derived on at least one line. This is the reason why both conditional lines are unmarked at this stage.

Because markings may come and go, we need a static definition to determine which derived formulas are the actual adaptive logic consequences.

Definition 26 φ is *finally derived* from T on line l of a finite stage s iff (i) φ is the second element of line l, (ii) line l is not marked at stage s, and (iii) every extension of the stage in which line l is marked may be further extended in such a way that line l is unmarked again.

This gives us the syntactic consequence relations **ALF** and **ALP**.

Definition 27 $T \vdash_{\mathbf{ALF}} \varphi$ (φ is finally **ALF**-derivable from T) iff φ is finally derived on a line of an **ALF**-proof from Γ.

Definition 28 $T \vdash_{\mathbf{ALP}} \varphi$ (φ is finally **ALP**-derivable from T) iff φ is finally derived on a line of an **ALP**-proof from Γ.

3.5 *Adaptive Proofs: Example*

We revisit the example of Sects. 2.8 and 2.10. We need to convert our theory $\langle T, F\rangle$ to the appropriate premise set for the adaptive proofs (in accordance with Theorem 2):

$$\{\circ(\psi \wedge \bigwedge F(\psi)) \mid \psi \in insts(T)\}.$$

For our example thus becomes

$$\{\circ\varphi_{i,j} \mid 1 \leq i \leq 5, j \in \{1, 2\}\}.$$

We abbreviate $\circ\varphi \wedge \neg\varphi$ to $!\varphi$.

Let us look in detail at Adaptive proof 1 (see Table 1). Lines 1–10 introduce the premises. All instances of the axioms of our theory are here introduced in such a way that they are not directly usable for further derivations. Lines 11–20 make them directly available. Of course this happens in a conditional way, i.e. the third element of the lines is no longer empty; we need to presuppose that its members are false in order for the second element to be derivable. On line 22 we have used both $\varphi_{3,1}$ and $\varphi_{2,1}$ for the derivation of $\neg R(e_1)$ and so we need to presuppose that both $!\varphi_{3,1}$ and $!\varphi_{2,1}$ are false. On line 24 we have used both $\varphi_{4,1}$ and $\varphi_{1,1}$ for the derivation of $R(e_1)$ and so we need to presuppose that both $!\varphi_{4,1}$ and $!\varphi_{1,1}$ are false.

The results of lines 22 and 24 contradict each other. This means that their presuppositions cannot be true together, i.e. $!\varphi_{3,1}$, $!\varphi_{2,1}$, $!\varphi_{4,1}$, and $!\varphi_{1,1}$ cannot be false together. This is made explicit on line 25. Line 25 teaches us that there is something wrong with the presuppositions that resp. $!\varphi_{3,1}$, $!\varphi_{2,1}$, $!\varphi_{4,1}$, and $!\varphi_{1,1}$ are false, so we need to mark all lines that use any of these presuppositions. The formulas on these marked lines are no longer considered ‘derived’.

Table 1 Adaptive proof 1: The flat case

1	$\circ\varphi_{1,1}$	$\emptyset$	PREM	
2	$\circ\varphi_{2,1}$	$\emptyset$	PREM	
3	$\circ\varphi_{3,1}$	$\emptyset$	PREM	
4	$\circ\varphi_{4,1}$	$\emptyset$	PREM	
5	$\circ\varphi_{5,1}$	$\emptyset$	PREM	
6	$\circ\varphi_{1,2}$	$\emptyset$	PREM	
7	$\circ\varphi_{2,2}$	$\emptyset$	PREM	
8	$\circ\varphi_{3,2}$	$\emptyset$	PREM	
9	$\circ\varphi_{4,2}$	$\emptyset$	PREM	
10	$\circ\varphi_{5,2}$	$\emptyset$	PREM	
11	$\varphi_{1,1}$	$\{!\varphi_{1,1}\}$	RD; 1	$\surd_{25}$
12	$\varphi_{2,1}$	$\{!\varphi_{2,1}\}$	RD; 2	$\surd_{25}$
13	$\varphi_{3,1}$	$\{!\varphi_{3,1}\}$	RD; 3	$\surd_{25}$
14	$\varphi_{4,1}$	$\{!\varphi_{4,1}\}$	RD; 4	$\surd_{25}$
15	$\varphi_{5,1}$	$\{!\varphi_{5,1}\}$	RD; 5	
16	$\varphi_{1,2}$	$\{!\varphi_{1,2}\}$	RD; 6	
17	$\varphi_{2,2}$	$\{!\varphi_{2,2}\}$	RD; 7	
18	$\varphi_{3,2}$	$\{!\varphi_{3,2}\}$	RD; 8	
19	$\varphi_{4,2}$	$\{!\varphi_{4,2}\}$	RD; 9	
20	$\varphi_{5,2}$	$\{!\varphi_{5,2}\}$	RD; 10	
21	$\neg L(e_1)$	$\{!\varphi_{3,1}\}$	RU; 13	$\surd_{25}$
22	$\neg R(e_1)$	$\{!\varphi_{3,1}, !\varphi_{2,1}\}$	RU; 21, 12	$\surd_{25}$
23	$C(e_1)$	$\{!\varphi_{4,1}\}$	RU; 14	$\surd_{25}$
24	$R(e_1)$	$\{!\varphi_{4,1}, !\varphi_{1,1}\}$	RU; 23, 11	$\surd_{25}$
25	$!\varphi_{1,1} \vee !\varphi_{2,1} \vee !\varphi_{3,1} \vee !\varphi_{4,1}$	$\emptyset$	MR; 22, 24	
26	$C(e_2)$	$\{!\varphi_{5,2}\}$	RU; 20	
27	$R(e_2)$	$\{!\varphi_{5,2}, !\varphi_{1,2}\}$	RU; 26, 16	
28	$\neg L(e_1) \vee R(e_1)$	$\{!\varphi_{3,1}\}$	RU; 21	
29	$\neg L(e_1) \vee R(e_1)$	$\{!\varphi_{4,1}, !\varphi_{1,1}\}$	RU; 24	

This does not mean that all conditional derivations get refuted. There is no problem with lines 15–20. So we can use these presuppositions to derive other unproblematic results on line 26 and 27. The subtle way in which the marking rule works has a, maybe surprising, effect. While line 21 and line 24 get marked remorselessly, the disjunction of the results of these lines, as derived on lines 28 and 29, does not get marked. The reason for this is a bit technical, but not difficult. We need to calculate the minimal choice sets of all minimal disjunctions of abnormalities. In this case the minimal choice sets are $\{!\varphi_{1,1}\}$, $\{!\varphi_{2,1}\}$, $\{!\varphi_{3,1}\}$, and $\{!\varphi_{4,1}\}$. Put in a slightly simplified way: we do not need to mark a line if its formula is derived on such conditions (third elements) that, for every choice set, there is a condition which does not overlap with the choice set. This is the case here:

$$\begin{aligned}
&\{!\varphi_{1,1}\} \cap \{!\varphi_{3,1}\} = \emptyset, \\
&\{!\varphi_{2,1}\} \cap \{!\varphi_{3,1}\} = \emptyset, \\
&\{!\varphi_{3,1}\} \cap \{!\varphi_{4,1}, !\varphi_{2,1}\} = \emptyset, \text{ and} \\
&\{!\varphi_{4,1}\} \cap \{!\varphi_{3,1}\} = \emptyset.
\end{aligned}$$

All markings in this proof are in some sense stable. We cannot derive any other or more minimal disjunctions of abnormalities. Neither can we derive the formulas of the lines on other conditions in such a way that they become unmarked. This means that all formulas on unmarked lines obtained here are finally derivable. So those formulas are all flat adaptive logic consequences of the premises.

Let us turn now to the prioritized adaptive proof (see Table 2). Now we have two kinds of abnormalities: those of the form $\circ^1\varphi \wedge \neg\varphi$ and those of the form $\circ^2\varphi \wedge \neg\varphi$. We use the following abbreviation:

$$!_i\varphi = \circ^i\varphi \wedge \neg\varphi.$$

The prioritized theory is translated to the premise set as follows:

$$\{\circ^1\varphi_{i,j} \mid 2 \leq i \leq 5, j \in \{1,2\}\} \cup \{\circ^2\varphi_{1,j} \mid j \in \{1,2\}\}.$$

Let us discuss the Adaptive proof 2 a bit. The only difference between the flat and the prioritized case is the way in which minimal choice sets are calculated. Given the prioritization the only minimal choice set (from line 25 on) is $\{!_2\varphi_{1,1}\}$. This means that we need to mark at most lines with $!_2\varphi_{1,1}$ as condition. The effect of this is that lines 2–4 and 21–23 stay 'derived' in the presence of line 25. $\neg R(e_1)$ is thus derived in this proof, unlike in the proof for the flat case. Moreover it is finally derived in this proof, i.e. every extension of this proof which marks line 22 can be extended to a proof which unmarks this line. So $\neg R(e_1)$ is a prioritized adaptive logic consequence of the premises.

4 Conclusion

A classically inconsistent theory T does not have any classical models. Hence, classical logic does not allow us to use such a theory in a meaningful way. This result does not accord with scientific practice, where the observation of a classical inconsistency does not always lead us to abandoning an axiomatic theory. How then do we use and understand classically inconsistent theories? The present analysis of paraconsistent reasoning is founded upon two ideas. First, we understand a classically inconsistent

Table 2 Adaptive proof 2: The prioritized case

1	$\circ^2\varphi_{1,1}$	$\emptyset$	PREM	
2	$\circ^1\varphi_{2,1}$	$\emptyset$	PREM	
3	$\circ^1\varphi_{3,1}$	$\emptyset$	PREM	
4	$\circ^1\varphi_{4,1}$	$\emptyset$	PREM	
5	$\circ^1\varphi_{5,1}$	$\emptyset$	PREM	
6	$\circ^2\varphi_{1,2}$	$\emptyset$	PREM	
7	$\circ^1\varphi_{2,2}$	$\emptyset$	PREM	
8	$\circ^1\varphi_{3,2}$	$\emptyset$	PREM	
9	$\circ^1\varphi_{4,2}$	$\emptyset$	PREM	
10	$\circ^1\varphi_{5,2}$	$\emptyset$	PREM	
11	$\varphi_{1,1}$	$\{!_2\varphi_{1,1}\}$	RD; 1	$\checkmark_{25}$
12	$\varphi_{2,1}$	$\{!_1\varphi_{2,1}\}$	RD; 2	
13	$\varphi_{3,1}$	$\{!_1\varphi_{3,1}\}$	RD; 3	
14	$\varphi_{4,1}$	$\{!_1\varphi_{4,1}\}$	RD; 4	
15	$\varphi_{5,1}$	$\{!_1\varphi_{5,1}\}$	RD; 5	
16	$\varphi_{1,2}$	$\{!_2\varphi_{1,2}\}$	RD; 6	
17	$\varphi_{2,2}$	$\{!_1\varphi_{2,2}\}$	RD; 7	
18	$\varphi_{3,2}$	$\{!_1\varphi_{3,2}\}$	RD; 8	
19	$\varphi_{4,2}$	$\{!_1\varphi_{4,2}\}$	RD; 9	
20	$\varphi_{5,2}$	$\{!_1\varphi_{5,2}\}$	RD; 10	
21	$\neg L(e_1)$	$\{!_1\varphi_{3,1}\}$	RU; 13	
22	$\neg R(e_1)$	$\{!_1\varphi_{3,1}, !_1\varphi_{2,1}\}$	RU; 21, 12	
23	$C(e_1)$	$\{!_1\varphi_{4,1}\}$	RU; 14	
24	$R(e_1)$	$\{!_1\varphi_{4,1}, !_2\varphi_{1,1}\}$	RU; 23, 11	$\checkmark_{25}$
25	$!_2\varphi_{1,1} \vee !_1\varphi_{2,1} \vee !_1\varphi_{3,1} \vee !_1\varphi_{4,1}$	$\emptyset$	MR; 22, 24	
26	$C(e_2)$	$\{!_1\varphi_{5,2}\}$	RU; 20	
27	$R(e_2)$	$\{!\varphi_{5,2}, !_1\varphi_{1,2}\}$	RU; 26, 16	
28	$\neg L(e_1) \vee R(e_1)$	$\{!_1\varphi_{3,1}\}$	RU; 21	
29	$\neg L(e_1) \vee R(e_1)$	$\{!_1\varphi_{4,1}, !_2\varphi_{1,1}\}$	RU; 24	$\checkmark_{25}$

theory T in such a way that the axioms of T are satisfied to a maximal extent. Second, this maximality condition can be spelled out using a modular semantics, in which the applications of T's axioms are considered as relatively independent and separable units. On the basis of these two ideas, we have developed a preferred models semantics of paraconsistent reasoning in science. This semantics respectively defines an inference relation for flat and prioritized axiomatic theories.

In the second part, we have provided these inference relations with a dynamic proof theory. In order to do this, we first expressed the inference relations in pure syntactic terms. Next, we have presented a flat and a lexicographic adaptive logic which we have proven to yield exactly the same results as the syntactic counterparts of the respective inference relations. Because the adaptive logics belong to the category of standard (lexicographic) adaptive logics, the adaptive characterization immediately gives rise to an adequate dynamic proof theory for the inference relations. We have concluded this paper by presenting and explaining this proof theory for both inference relations (flat and prioritized).

References

1. Andreas, H. (201x). Networks of partial structures. under review.
2. Balzer, W., Moulines, C. U., & Sneed, J. (1987). *An architectonic for science. The structuralist program*. Dordrecht: D. Reidel Publishing Company.
3. Batens, D. (2007). A universal logic approach to adaptive logics. *Logica Universalis*, *1*, 221–242.
4. Benferhat, S., Dubois, D., & Prade, H. (1997). Some syntactic approaches to the handling of inconsistent knowledge bases: A comparative study part 1: the flat case. *Studia Logica*, *58*(1), 17–45.
5. Brewka, G. (1991). Belief revision in a framework for default reasoning. In *Proceedings of the Workshop on The Logic of Theory Change* (pp. 602–622). London: Springer.
6. Brown, B., & Priest, G. (2004). Chunk and permeate: a paraconsistent inference strategy. part 1: the infinitesimal calculus. *Journal of Philosophical Logic*, *33*, 379–388.
7. da Costa, N., & French, S. (2003). *Science and partial truth*. Oxford: Oxford University Press.
8. Frisch, M. (2005). *Inconsistency, asymmetry, and non-locality*. Oxford: Oxford University Press.
9. Hansson, S. O. (1999). *A textbook of belief dynamics. Theory change and database updating*. Dordrecht: Kluwer.
10. Kraus, S., Lehmann, D., & Madigor, M. (1990). Nonmonotonic reasoning, preferential models and cumulative logics. *Artificial Intelligence*, *44*, 167–207.
11. Meheus, J., Straßer, C., & Verdée, P. (2016). Which style of reasoning to choose in the face of conflicting information? *Journal of Logic and Computation*, *26*(1), 361–380. doi:10.1093/logcom/ext030.
12. Putte, F. V. D. & Straßer, C. (2012). Extending the standard format of adaptive logics to the prioritized case. *Logique Et Analyse*, *220*, 601–641.
13. Rescher, N., & Manor, R. (1970). On inference from inconsistent premisses. *Theory and Decision*, *1*(2), 179–217.
14. Shoham, Y. (1988). *Reasoning about change: Time and causation from the standpoint of artificial intelligence*. Cambridge, MA: MIT Press.
15. Sneed, J. (1979). *The logical structure of mathematical physics*. Dordrecht: D. Reidel Publishing Company.
16. Straßer, C. (2014). *Adaptive logics for defeasible reasoning* (Vol. 38). Trends in Logic Cham: Springer. *Applications in Argumentation, Normative Reasoning and Default Reasoning*, Heidelberg.
17. Wagner, G. (1991). Ex contradictione nihil sequitur. In *Proceedings of the 12th IJCAI* (pp. 538–543), Sidney, Australia.

Inconsistency in *Ceteris Paribus* Imagination

Francesco Berto

Abstract I propose to model imagination as a ceteris paribus modal operator: a variably strict world quantifier in a modal framework including both possible and so-called non-normal or impossible worlds. The latter secure lack of closure under classical logical consequence for the relevant mental states, while the variability of strictness captures how the agent imports information from actuality in the imagined non-actual scenarios. The proposed formal semantics models how a conceiving agent can imagine inconsistencies. I also discuss how similarity may work when impossible worlds are around.

1 Intro: Imagination as a Modal

"Imagining" as well as "conceiving" refer in this paper to a range of intentional phenomena. Intentionality is the feature of those mental states that are directed to, and involve the representation of, objects and configurations thereof, situations, or circumstances. Chalmers [6] characterizes a notion he calls *positive conceivability*: when we positively conceive that p, we do not just assume or suppose that p, as when we make an assumption in a mathematical proof. Rather, we represent a sce-

The research leading to this paper has been funded within the 2013–15 AHRC project *The Metaphysical basis of Logic: the Law of Non-Contradiction as Basic Knowledge* (grant ref. AH/K001698/1). The technical parts draw on a paper published in *Proceedings of The Aristotelian Society*, 114(2014): 103–21. I am very grateful to the Editors for allowing me to re-use that material. Parts of the work have been presented in 2014 at the conference *Paraconsistent Reasoning in Science and Mathematics* at the Munich Center for Mathematical Philosophy, at the Modality Seminar at the University of Aberdeen, at the Groningen Logic Colloquium (GroLog) at the University of Groningen, and at the *Grandes Conférences des Archives Poincaré* in Nancy. I am grateful to all the audiences for the many helpful remarks and comments.

F. Berto (✉)
Department of Philosophy and Institute for Logic, Language and Computation (ILLC), University of Amsterdam, Amsterdam, Netherlands
e-mail: F.Berto@uva.nl

H. Andreas and P. Verdée (eds.), *Logical Studies of Paraconsistent Reasoning in Science and Mathematics*, Trends in Logic 45,
DOI 10.1007/978-3-319-40220-8_3

nario – a configuration of objects and properties – correctly described by p, in our imagination.[1]

That we have such intentional states is first-hand experience for us: the human mind has the ability to conceive or imagine rich and detailed alternatives to actuality in order to extract information from them. And this has a very pragmatic motivation. We cannot experience beforehand which scenarios are or will be actual for us to face in real life. So we explore them in our mind, leaving our perceptions "offline": How will the financial markets react if Russia defaults? What contingency plans would you adopt if you failed your logic class? Would Mr. Jones show the symptoms he shows, had he taken aspirin? A vast literature on "counterfactual imagination" in cognitive science [5, 15] shows how such mental activity improves our performances.

The *logical* study of intentionality flourished when authors such as Hintikka [13] realized that the modal framework of possible worlds semantics could be applied to the analysis of intentional states like knowledge, belief, psychological information. This was one of the successes of philosophical logic, whose results were taken up by linguistics and became influential in computer science and Artificial Intelligence (see Fagin et al. [11], Meyer and van der Hoek [20]). The key insight was: cognitive agent x ®s that p, with ® the relevant representational mental state (knows, believes, is informed that), when p holds throughout a set of worlds: those compatible with x's evidence, overall beliefs, etc. Accessibility relations single out these worlds: the accessible worlds are the scenarios x entertains. Let R be one such accessibility: "wRw_1" means "World w_1 is an epistemic alternative for world w". Read "®p" as "It is represented [believed, known, etc.] that p". Then the (non-agent-indexed) truth conditions for ® are ("iff" = "if and only if"):

'®p' is true at w iff p is true at all w_1, such that wRw_1.

Various logical works have applied this framework specifically to the treatment of imagination as a modal operator [8, 21]. However, in ordinary epistemic logics the modelling of such mental states via possible worlds semantics originates the so-called "problem of logical omniscience", whereby the relevant cognitive agents are inevitably modelled as logically idealised. Mental states come out closed under logical consequence or entailment:

(*Closure*) If ®p, and p entails q, then ®q.

That is, agents represent (know, believe, imagine, etc.) all the logical consequences of what they represent (they are "logically omniscient" in this sense). As a special case, all logically valid formulae are represented:

(*Validity*) If p is valid, then ®p.

[1] While rationalists like Descartes made a lot of a distinction between conceiving and imagining, empiricists like Hume blurred it. Given the aforesaid rough characterisation of the mental act of representing a scenario, we can use "conceiving" and "imagining" broadly as synonyms. In particular, the imagined scenarios need not be *visually* imaginable; for instance, they may involve abstract objects.

And mental states are perforce consistent:

(*Consistency*) $\sim(\circledR p \wedge \circledR{\sim}p)$.

Such principles hold in the weakest normal modal logic K (for Consistency, we just add the D-principle: accessibility is serial). They follow from interpreting epistemic operators as quantifiers on *possible* (logically closed, consistent) worlds. And it is universally admitted (e.g. Meyer and van der Hoek [20], Sect. 2.5) that they deliver implausibly idealized mental states. We experience having (perhaps covert) inconsistent beliefs. Excluded Middle is (let us suppose) valid, but intuitionists do not believe it. We know basic arithmetic truths like Peano's postulates; and these entail (let us suppose) Goldbach's conjecture; but we don't know whether Goldbach's conjecture is true. The cognitive agency so modelled has little to do with human intelligence.

Logical omniscience is obviously connected to the topic of *hyperintensionality.* A concept is hyperintensional when it draws a distinction between two intensionally or necessarily equivalent contents. Now intentional states do draw distinctions between intensionally (necessarily) equivalent contents: $\circledR p$ may differ from $\circledR q$ even when p and q are logically equivalent. But the possible worlds apparatus can only draw intensional, not hyperintensional, distinctions. So it is unsuitable for the modelling of conceivability and connected doxastic and informational notions. Different approaches to hyperintensionality have been proposed in the logical literature – for instance, Tichy's Transparent Intensional Logic (Duží, Jespersen and Materna [10]), or structuralist approaches to content (King [16]). Each is promising and each faces troubles (see e.g. Ripley [26], Jago [14], for a set of thorough objections to structuralism).

This paper relies on a different approach to hyperintensional mental states, and specifically to imagination – an approach which has the feature of preserving the modelling of such states as restricted quantifiers on worlds, thereby taking advantage of the core idea of worlds semantics.

2 Impossible Worlds

The approach, pursued in epistemic logic by Rantala [25] and developed by other authors such as Priest [23], expands the world machinery by adding *non-normal* or *impossible* worlds (see Berto [3] for an overview; Berto [2]) and, again, Priest [23], for applications to imagined fictional and nonexistent objects). If possible worlds are ways things could be, then impossible worlds are ways things could *not* be: they represent absolute (logical, mathematical, metaphysical) impossibilities, such as contradictions, as obtaining. In epistemic logic, non-normal worlds are understood as viable epistemic alternatives from the viewpoint of imperfect or inconsistent cognitive agents. Wansing [28] convincingly showed how non-normal worlds semantics can provide a very general framework for epistemic logics. The relevant intentional operators are still interpreted, as in the standard approach, as modals – as restricted quantifiers on worlds. But they are now quantifiers on non-normal worlds as well.

By accessing such worlds in the truth conditions of the relevant ®, one easily refutes Validity, Closure and Consistency. E.g. for Closure: take a non-normal world w where p holds, but $p \vee q$ fails. If w is accessible (to the relevant agent), we have ®p without ®($p \vee q$), although p logically entails $p \vee q$. For Consistency: just access a non-normal world where both p and $\sim p$ hold to get ®p and ®$\sim p$.

In this paper I want to show how impossible worlds also help with a problem, which is in a sense symmetric to the issue of logical omniscience. On the one hand, our representational mental states should sometimes be inconsistent, and/or not closed under entailment: we do not imagine everything that follows from what we explicitly conceive, and we can occasionally have inconsistent conceptions. This is precisely the logical omniscience issue. But on the other hand, we do conceive or imagine things *not* logically entailed by what is explicitly included in the mental act of imagining a scenario.

What does "explicit" mean here? It seems plausible to say that, when we engage in a conscious act of imagination whereby we conceive a non-actual scenario, such an act has some deliberate basis, whereby we purposefully focus on a given content. Let us call such content *explicit*. A simple example: conceiving subject x reads one of Arthur Conan Doyle's novels, portraying Sherlock Holmes as a man who is variously active in London, so-and-so dressed, doing this and that. On the basis of the input overtly given in the text, x starts forming a mental representation of the situation described there.

However, when we engage in such exercises of imagination, we typically do not limit ourselves to the information which we explicitly represent in our minds, or to what follows from it logically, and there is an obvious sense in which we do it legitimately. Sticking to our example: in reality, London is in the UK and normally endowed men have kidneys, although Doyle's stories (assume) do not claim this explicitly. Now x can take such information as holding throughout the represented situation, absent information to the contrary: x does imagine Holmes as a normally endowed man with kidneys and as living in the UK. This integration is typical: we do not conceive such additional details by inferring them logically from the explicitly given content, rather by *importing* background information we already have, and which we retain in the non-actual scenario we build a mental representation of.

If this is right, then such exercises of imagination we do both less and more than applying a fixed set of logical rules of inference: we do less, because we don't draw all the logical consequences of what we explicitly conceive. But we also do more, because we develop our imagined scenarios by importing what does not follow logically from their explicit content. As Timothy Williamson has claimed in *The Philosophy of Philosophy*, we should then avoid looking for smooth logical rules governing such exercises:

> Calling [the relevant conceiving] "inferential" is no longer very informative. [...] To call the new judgment "inferential" simply because it is not made independently of all the thinker's prior beliefs or suppositions is to stretch the term "inferential" beyond its useful span. At any rate, the judgment cannot be derived from the prior beliefs or suppositions purely by the application of general rules of inference. (Williamson [31], pp. 147 and 151)

Our imagining or positively conceiving non-actual scenarios is based, rather, on imaginative simulation. I think that impossible worlds can help with what is going on here as well.

3 Imagination as *Ceteris Paribus* Activity

I propose to model imagination, so understood, via modal operators interpreted as *variably strict* quantifiers on worlds, possible and impossible. The addition of impossible worlds has the role of accounting for our capacity of imagining or conceiving absolute impossibilities and inconsistencies, and for the lack of logical closure for our imagined scenarios. The variability of strictness is to account for the (highly contextual) selection of the information we import in a representational mental act when we integrate its explicit content.

The explicit content itself may play a role similar to that of the antecedent of a *variably strict conditional*. These conditionals, which can be indicative or subjunctive-counterfactual, are sometimes called also "non-monotonic conditionals", or "dependent conditionals" (Bennett [1], pp. 16–17). They are such that, as Bennett says, "the consequent is reachable from the antecedent only with help from unstated particular matters of fact" (Ibid.). They have been called "*ceteris paribus* conditionals" in the Chapter on conditional logics of Priest's nowadays popular *Introduction to Non-Classical Logic,* because they embed an implicit "other things being equal" clause:

> We can say: "if it does not rain tomorrow, then, other things being equal, we will go to the cricket", or "if it does not rain tomorrow and everything else relevant remains unchanged, we will go to the cricket". The Latin for "other things being equal" is *ceteris paribus*, so we can call this a *ceteris paribus* clause. (Priest [24], p. 84)

I will extend the terminology to our imagined operators, and will thereby talk of "*ceteris paribus* imagination". What is actually imagined in exercises of imagination of this kind, is what holds in worlds where the antecedent holds *and* further information imported from actuality holds, too. An accessible world is one that complies with such explicit content, but additionally, does not bring gratuitous changes with respect to how we take the actual world to be. We will expand on this initial characterization in the following.

A similar framework is proposed in Lewis' [18] famous paper on truth in fiction, whose key idea is that "we can help ourselves to the notion of what is *explicitly* so according to the fiction and use the notion of possible worlds to extend outwards and define what is *implicitly* so" (Sainsbury [27], p. 76). The explicit fictional content corresponds to the explicit content of our imagined scenarios, and works, in Lewis' approach, too, like the antecedent of a *ceteris paribus* conditional. As Williamson also claims:

> We seem to have a prereflective tendency to minimum alteration in imagining counterfactual alternatives to actuality, reminiscent of the role that similarity between possible worlds plays in the Lewis–Stalnaker semantics. (Williamson [31], p. 151)

We will come back to the topic of world *similarity*, and to the question whether such imaginative exercises come closer to *counterfactual* or subjunctive, or rather to indicative, *ceteris paribus* conditionals, later on in this paper.[2] For now, let us just remark that for Williamson *ceteris paribus* exercises of imagination are what we engage in when we evaluate the relevant conditionals in our daily life: we first explicitly imagine the antecedent, then we develop our imagined scenario to see whether such an expansion eventually leads us to verify the consequent.

Let us now begin to make the idea of *ceteris paribus* imagination more precise.

4 A Semantics for Imagination

Let L be a sentential language with atoms p, q, r ($p_1, p_2, \ldots, p_n$), negation $\sim$, conjunction $\wedge$, disjunction $\vee$, the conditional $\rightarrow$, the standard modals $\Box$ and $\Diamond$, square and round brackets [and], (and). While the round brackets are the usual auxiliary symbols preventing scope ambiguities, the square brackets allow forming variably strict modal operators, which can then be prefixed to formulas. The well-formed formulas of L are the atoms and, if A and B are formulas, $\sim A$, $(A \wedge B)$, $(A \vee B)$, $(A \rightarrow B)$, $\Box A$, $\Diamond A$, and $[A]B$ (outermost brackets are usually omitted).

The only piece of notational novelty is $[A]$, to be thought of as a sententially indexed modal operator. Consider a bunch of *acts of imagining* performed by a given cognitive agent on specific occasions, and suppose each is characterized by its explicit content, to be directly expressed by a formula of L. Let the set of such formulas be K. Each $A \in$ K determines its own operator, $[A]$, and its own accessibility relation, R_A (the idea, in the context of conditional logics, goes back to Chellas [7]). One can read "$[A]B$" as: "It is imagined in act A that B"; or, less tersely and more accurately: "It is imagined in the act whose explicit content is A, that B".[3]

An interpretation for L is a sextuple $< P, I, @, \{R_A | A \in K\}, \Vdash >$. P is the set of possible worlds; I is the set of impossible worlds. P and I are disjoint, W = P $\cup$ I is the totality of worlds. @ $\in$ P is the actual world. $\{R_A | A \in K\}$ is a set of binary accessibilities on W, i.e., $R_A \subseteq$ W x W. $\Vdash$ is a pair $< \Vdash^+, \Vdash^- >$ of relations between worlds and formulas: "$w \Vdash^+ A$" says that A is true at world w, "$w \Vdash^- A$" says that A is false there.

Truth and falsity conditions will be given separately: as we want to model imaginable inconsistencies, we allow some formulas to be both true and false, or "glutty", at some worlds (and also, neither true nor false, or "gappy"). However, one may not want this to happen at possible worlds: their being consistent and maximal is what

[2]As remarked by many, including Lewis [17] and Williamson [31], counterfactuals are ill-named if there are, as can be argued, meaningful and true (though, admittedly, pragmatically odd to assert) "counterfactuals" with true antecedent. So some people prefer to talk of "subjunctives", e.g. Bennett [1]. Both kinds of terminology are common in the literature anyway.

[3]Intentional operators are often indexed to agents: in the usual notation of epistemic logics, "$K_x A$" is to mean that cognitive agent x knows/believes that A. Since the subscript would not have done much work in our essentially single-agent setting, I have omitted it.

makes them *possible*, one may say. To accommodate this view, we can restrict our attention to those interpretations of L which begin by relating atomic formulas at worlds to truth, falsity, both, or neither, by complying with a Classicality Condition on *possible* worlds:

(CC) If $w \in$ P, then for each p, either $w \Vdash^{+} p$ or $w \Vdash^{-} p$, but not both.

Next, the truth and falsity conditions are defined for all $w \in$ P, that is, for *possible* worlds, as follows. The extensional operators works thus:

$$w \Vdash^{+} \sim A \text{ iff } w \Vdash^{-} A$$
$$w \Vdash^{-} \sim A \text{ iff } w \Vdash^{+} A$$
$$w \Vdash^{+} A \wedge B \text{ iff } w \Vdash^{+} A \text{ and } w \Vdash^{+} B$$
$$w \Vdash^{-} A \wedge B \text{ iff } w \Vdash^{-} A \text{ and } w \Vdash^{-} B$$
$$w \Vdash^{+} A \vee B \text{ iff } w \Vdash^{+} A \text{ or } w \Vdash^{+} B$$
$$w \Vdash^{-} A \vee B \text{ iff } w \Vdash^{-} A \text{ or } w \Vdash^{-} B$$

As for the modal operators:

$$w \Vdash^{+} \Box A \text{ iff for all } w_1 \in \text{P}, w_1 \Vdash^{+} A$$
$$w \Vdash^{-} \Box A \text{ iff for some } w_1 \in \text{P}, w_1 \Vdash^{-} A$$
$$w \Vdash^{+} \Diamond A \text{ iff for some } w_1 \in \text{P}, w_1 \Vdash^{+} A$$
$$w \Vdash^{-} \Diamond A \text{ iff for all } w_1 \in \text{P}, w_1 \Vdash^{-} A$$

Unrestricted necessity/possibility is defined at possible worlds as truth at all/some possible world(s). Our conditional is strict:

$$w \Vdash^{+} A \rightarrow B \text{ iff for all } w_1 \in \text{P}, \text{ if } w_1 \Vdash^{+} A, \text{ then } w_1 \Vdash^{+} B.$$
$$w \Vdash^{-} A \rightarrow B \text{ iff for some } w_1 \in \text{P}, w_1 \Vdash^{+} A \text{ and } w_1 \Vdash^{-} B.$$

An easy induction on the complexity of formulas will show that the CC generalizes: no gaps or gluts at possible worlds for any formula composed of the logical vocabulary defined so far. And, so far, we have just propositional S5, except for the peculiarity of spelling out truth and falsity conditions separately. However, at points in I all complex formulas are treated as atomic, that is, truth values are assigned to them *directly*, not recursively (this is the original trick of Rantala [25]): a conjunction may be true there even though one of the conjuncts is false, etc. Impossible worlds, thus, can fail to be closed under any non-trivial relation of logical consequence. Now the key move is to allow access to such anarchic worlds via our indexed accessibilities. The truth and falsity conditions for $[A]$, again for $w \in$ P, are:

$$w \Vdash^{+} [A]B \text{ iff for all } w_1 \in \text{W such that } wR_A w_1, w_1 \Vdash^{+} B$$
$$w \Vdash^{-} [A]B \text{ iff for some } w_1 \in \text{W such that } wR_A w_1, w_1 \Vdash^{-} B$$

(At impossible worlds, formulas of the form $[A]B$ may be assigned arbitrary values). We understand "$wR_A w_1$" as saying that w_1 realizes an intentional state: things are there as they are represented (at w) in the act of imagination whose explicit content is expressed by A. In this respect, $[A]B$ is precisely similar to a *ceteris paribus* conditional where A works as the antecedent. But while in the Lewis–Stalnaker approach only possible worlds show up in the semantics, and thus any such conditional with an antecedent true at no possible world is automatically true, here we have at our disposal, and we can access, impossible worlds as well. This will allow, as we shall see, some of the hyperintensional distinctions we wanted acts of imagination to be able to make.

The two clauses above may be equivalently formulated using set-selection functions, as also Lewis does in *Counterfactuals* (see Lewis [17], pp. 57–9). Each formula A in K comes with a function, f_A, which takes as input the world w where the act of imagination takes place, and selects the set of worlds accessed via that act: $f_A(w) = \{w_1 \in W | wR_A w_1\}$. Taking $|A|$ as the set of worlds making A true, we get, for $w \in$ P:

$$w \Vdash^+ [A]B \text{ iff } f_A(w) \subseteq |B|$$
$$w \Vdash^- [A]B \text{ iff } f_A(w) \cap |{\sim}B| \neq \emptyset$$

So $[A]B$ is true (false) at w iff B is true at all worlds (false at some world) in a set selected by f_A.[4] Since $wR_A w_1$ if and only if $w_1 \in f_A(w)$, the two formulations are interchangeable.

Finally, logical consequence gets the straightforward definition for modal frames with a designated base world. Where S is a set of formulas:

$$S \models A \text{ iff, in every interpretation } <\text{P, I, @}, \{R_A | A \in \text{K}\}, \Vdash>, \text{ if @} \Vdash^+ B \text{ for all } B \in S, \text{ then @} \Vdash^+ A.$$

Logical consequence is truth preservation at the actual world in all interpretations (as a special case, logical validity is truth at the actual world in all interpretations: $\models A$ iff $\emptyset \models A$). What matters for our purposes is that @ $\in$ P, so we define consequence by only looking at a world which is *possible*: "impossible worlds are only a figment of the agents' imagination: they serve only at epistemic alternatives. Thus, logical implication and validity are determined solely with respect to the standard worlds." (Fagin et al. [11], p. 358). If we understand impossible worlds as "worlds where

[4] A technical point. When the truth and falsity conditions are spelt thus, the CC does not generalize seamlessly to formulas comprising our *ceteris paribus* operators. Since the motivation for having the CC in the first place was to avoid gaps and gluts of truth values at possible worlds, this may be unsatisfactory: we don't want a merely imagined inconsistency, for instance, to generate a possible or actual one. In order to fix this so that the CC extends throughout the whole language, one would need, in fact, to rephrase the falsity conditions for our operators; to rule out gluts, e.g., one would need: $w \Vdash^- [A]B$ iff not $w \Vdash^+ [A]B$.

logic may be different", this looks like a natural move: we want to define logical consequence with respect to worlds where logic is not different.[5]

I should mention that a framework similar to this one can be found, albeit in a different context that does not aim at modelling imagination directly, in Wansing [29].[6] This paper gives an intuitively plausible modal semantics for connexive logics (see Wansing [30] for a general introduction), with truth and falsity conditions spelt out separately, and "dynamic" truth and falsity conditions for conditionals. Although the falsity conditions for Wansing's logic are different from the ones presented here, there are many interesting similarities worth exploring.

5 Constraints

The semantics for *ceteris paribus* conditionals in the Lewis–Stalnaker approach is based on a notion of closeness between worlds, which is understood as similarity. Roughly, a conditional of this kind, let us say, a subjunctive "If it were the case that A, then it would be the case that B", is true at world w iff the world(s) most similar to w where A holds also make true B. However, world similarity has been variously criticized as a desperately vague and context-dependent notion. While some attempts have been made to make the notion viable in Artificial Intelligence (Delgrande [9]), things seem to get worse when non-normal worlds, as worlds representing absolute impossibilities, are around. How does similarity work for *them*? Supposing mathematical truths are unrestrictedly necessary, is a world where the Axiom of Choice fails closer than one where Fermat's Last Theorem is false? In spite of some work (Nolan [22], Brogaard and Salerno [4]), this is a largely unexplored territory.[7]

However, we can explore the plausibility of at least some inferential schemas involving our *ceteris paribus* operators without taking a precise stance on world closeness as similarity. We can wonder whether we want some constraints on the various R_A's or f_A's to hold by looking at what goes on, in a similar setting, with the

---

[5]It is sometimes claimed that this kind of impossible worlds semantics alters the meaning of the logical vocabulary. This is connected to the understanding of non-normal worlds as points where
(Footnote 5 continued)
"logic may be different": for instance, a world where a contradiction, $p \wedge {\sim} p$, is true, one may say, is one where that formula does not express the proposition *that p and not-p*. However, the semantics above does validate the Law of Non-Contradiction, and furthermore (when fixed as per the previous footnote) provides no counterexamples to it. A world where a contradiction is true is a way things could *not* be according to the semantics, but a way things *could* be is someone's positively conceiving a contradiction, that is, imagining a scenario in which it obtains, and impossible worlds have the role of modelling such acts of imagination. Dialetheists like Graham Priest believe that the actual world is inconsistent, and it is controversial whether they are thereby automatically misunderstanding the meaning of negation (or that of conjunction).

[6] Thanks to an anonymous Referee for pointing me at Wansing's work.

[7]But Jago [14] develops a nice framework which allows a distinction between *obvious* and *subtle* logical impossibilities, via a total ordering of impossible worlds with respect to the degree of complexity of the logical truths violated at them. I suspect that Jago's techniques may be used to provide some kind of similarity (or at least of logical similarity) metric for impossible worlds.

ceteris paribus conditionals of weak conditional logics such as C^+ (see Priest [24], pp. 87–90); for these logics work without presupposing a metric for the closeness of worlds.

Here is one plausible basic constraint:

$$(\textit{Obtaining})\ \text{If } w \in \mathrm{P}, \text{ then } f_A(w) \subseteq |A|$$

Possible worlds only access worlds where the explicit content obtains. *Obtaining* gives this logical validity:

$$\models [A]A$$

It is obvious that one imagines what one explicitly imagines. Next, these entailments also look clearly right:

$$[A](B \wedge C) \models [A]B$$
$$[A](B \wedge C) \models [A]C$$

It seems obvious that, when I imagine that $B \wedge C$ is the case, I also imagine each conjunct. This is secured by the appropriate constraint:

(*Simplification*) If $w \in \mathrm{P}$, then if $wR_A w_1$ then (if $w_1 \Vdash^+ B \wedge C$, then $w_1 \Vdash^+ B$ and $w_1 \Vdash^+ C$).

The companion constraint:

(*Adjunction*) If $w \in \mathrm{P}$, then if $wR_A w_1$ then (if $w_1 \Vdash^+ B$ and $w_1 \Vdash^+ C$, then $w_1 \Vdash^+ B \wedge C$)

Gives us:

$$[A]B, [A]C \models [A](B \wedge C).$$

When both constraints are accepted, accessibility is limited to worlds which are, so to speak, fully well-mannered with respect to conjunction: they satisfy Conjunction Introduction as well as Conjunction Elimination. However, *Adjunction* might look controversial, for it may be doubted that, when one imagines in one act $[A]$ that B and that C, one automatically imagines that $B \wedge C$. A popular example due to Quine will allow us to discuss this. Quine's original story concerned counterfactual conditionals, and is usually taken as evidence for the role that context and the consequent play in their evaluation. But one can easily rephrase it in terms of *ceteris paribus* imagination.

The explicitly imagined situation, $[A]$, is one in which Caesar the Roman emperor is in command of the US troops in the Korean war. Given the same explicit content as input (say, a short science fiction story on time travel), we may imagine Caesar using the atomic bomb, B, or we may imagine him using catapults, C. We can imagine Caesar dropping the bomb, $[A]B$, as we import in the representation information concerning the weapons available in the Fifties. We can imagine him dropping stones to the Reds via catapults, $[A]C$, as we import the setting of the ancient Romans' military apparatus. But we would not infer that $[A](B \wedge C)$, we imagine Caesar employing both the bomb and catapults. We *can* imagine that as well, making the scenario even weirder, but that should not come as an automatic logical entailment.

However, I think that something has gone wrong in this reconstruction of the situation, though what has gone wrong is not detected by our current formalism. We are focusing on single acts of imagining, but we individuate them only via their explicit content. However, it seems plausible that different acts of imagining will trigger the importation of different background information depending on contexts (the time and place at which the cognitive agent performs the act, the status of its background information, etc.). And it seems clear that there is a shift in context in the Quinean example. So I think that Adjunction can be maintained on context-dependent grounds. The formalism may represent this by adding a set of contexts to the interpretations, variables ranging on them in the language, and by directly indexing representational acts with contexts: $[A]_x$, $[A]_y$, for instance, will stand for two distinct acts with the same explicit content, A, performed in contexts x and y. Once the adjunctive inference is parameterized to same-indexed contents, it should work fine.

To explore a plausible further constraint, start with a special case of $\models [A]A$, namely:

$$\models [A \wedge B](A \wedge B)$$

Via Simplification, we get:

$$\models [A \wedge B]A$$

Now a condition one *may* want to have is what we may call the Law of Imaginative Equivalents:

$$\text{(LIE) If } f_A(w) \subseteq |B| \text{ and } f_B(w) \subseteq |A|, \text{ then } f_A(w) = f_B(w).$$

If all the worlds selected by f_A make B true and vice versa, then A and B are equivalent in our imagination: when we imagine either, we look at the same set of scenarios. (LIE) validates this:

(*Substitutivity*) $[A]B, [B]A, [B]C \models [A]C$

This says that two imaginative equivalents A and B can be replaced *salva veritate* with each other within brackets. Suppose, for instance, that *house* and *habitation* are for you imaginative equivalents: you cannot imagine that something is a house without imagining that it is a habitation and vice versa. Therefore, $[A]B$: when you explicitly imagine that your home is a house, you imagine that it is a habitation; and $[B]A$: when you explicitly imagine that your home is a habitation, you imagine that it is a house. Suppose $[B]C$: as you imagine that your home is a habitation, you imagine that it looks nice. It follows that the same happens when you imagine that your home is a house.

(LIE) also validates an inference we may call Restricted Transitivity[8]:

$$\text{(RT) } [A]B, [A\wedge B]C \models [A]C$$

(This is a bit more difficult to see, so I will add a proof: suppose (1) $@ \Vdash^{+} [A]B$ and (2) $@ \Vdash^{+} [A\wedge B]C$. From (1) and $[A]A$ (secured by Obtaining), via Adjunction, we get $@ \Vdash^{+} [A](A\wedge B)$.$[A\wedge B]A$ is valid (from $\models [A\wedge B](A\wedge B)$, via Simplification), so $@ \Vdash^{+} [A\wedge B]A$. Applying the truth clause for [ ], to the last two we get $f_A(@) \subseteq |A \wedge B|$ and $f_{A\wedge B}(@) \subseteq |A|$. By (LIE), $f_A(@) = f_{A\wedge B}(@)$. From (2), by the truth clause for [ ] again, $f_{A\wedge B}(@) \subseteq |C|$. From this and the previous identity, $f_A(@) \subseteq |C|$. From this via the truth clause for [ ] again, $@ \Vdash^{+} [A]C$).

Ceteris paribus conditionals are, notoriously, "non-monotonic" in the sense that Antecedent Strengthening fails from them: a counterfactual "If it were the case that *A*, then it would be the case that *B*" does not entail "If it were the case that *A* and *C*, then it would be the case that *B*". Our imagination operators immediately inherit such a nice feature, in that the following inference is invalid in the semantics:

$$[A]B \models_? [A \wedge C]B$$

An act of imagination (in a given context) is individuated by its explicit content. But then one cannot automatically import further information into the explicit content itself without turning it into a different act. I imagine that I fail my logic class, and I will imagine myself in a sad mood. But if I imagine failing my logic class and that everyone else has failed, so that the exam needs to be re-taken with an easier array of exercises, my mood will not be that sad in such a scenario. The variability in the strictness of our operators is the essential tool securing such non-monotonicity of our exercises of imagination.

Next, there are some invalidities essentially involving the hyperintensional features of our operators. Here is one:

$$A \rightarrow B \models_? [A]B$$

Recall that the premise is an intensional (strict) conditional: all the possible *A*-worlds are *B*-worlds. However, in an act of imagination whose explicit content is given by *A*, we do not automatically imagine that *B*: as our act is hyperintensional, that is, it discriminates between various absolute impossibilities, we may look at impossible *A*-worlds where *B* fails. In particular, strict conditionals which logicians in the tradition of relevant logics (see Mares [19] for a nice introduction) call "irrelevant", such as conditionals which hold just because the antecedent is impossible, or the consequent necessary, do not imply the corresponding irrelevant conceivings. In our semantics, this is fine:

[8]General Transitivity fails for our operators, just as it does for *ceteris paribus* conditionals. Its failure is a consequence of the failure of Antecedent Strengthening, to which we are about to come.

$$\models (A \wedge \sim A) \rightarrow B.$$

However, this fails:

$$\models_? [A \wedge \sim A]B$$

(Just peek at an inconsistent $w \Vdash^+ p \wedge \sim p$ where, however, it is not the case that $w \Vdash^+ q$). That we explicitly imagine an inconsistent scenario does not mean that we trivialize our act of imagination. Similarly, although we have:

$$\models A \rightarrow (B \vee \sim B),$$

the corresponding inference concerning imagination fails:

$$\models_? [A](B \vee \sim B).$$

In general, we can discriminate between logical or absolute necessities and we do not conceive them automatically, independently of what we (explicitly) conceive. Thanks to impossible worlds, we have neither:

$$\Box B \models_? [A]B,$$

nor:

$$\sim \Diamond A \models_? [A]B.$$

Such failures make a decisive difference with respect to approaches to conditional logic that model *ceteris paribus* operators as sententially indexed modals but use only possible worlds, as in the aforementioned Chellas [7], and further highlight the necessity of using impossible worlds for our purposes. One can model *some* inconsistent conceptions using the Chellas framework, for instance $[A]B$ and $[A]{\sim}B$,[9] by looking at an empty set of worlds via the relevant accessibility. One cannot, however, avoid for instance trivialization when an inconsistency is explicitly conceived, i.e., when it shows up in our formalism as: $[A \wedge \sim\sim A] \ldots$ Then any formula can take the place of the dots because there is no world in which what is within brackets is true. However we don't want to imagine arbitrary contents just because we explicitly imagine an inconsistency. Or, just to give another example, we want to have $[A]B$ without having $[A](B \vee C)$: when we imagine in the act whose explicit content is A, that B, we do not thereby imagine a disjunction between B and an arbitrary C (one explicitly imagines Holmes in London and one imagines him in England, but it does not follow that one imagines that either Holmes is in England or Moriarty is on the Moon). The only way to achieve this is to have impossible worlds where disjunction misbehaves in such a way that B is true there while $B \vee C$ is not.

[9] As pointed out to me by an anonymous Referee.

6 Cotenability and Imaginative *Modus Ponens*

To discuss the plausibility of one further constraint, we need to introduce the notion of *cotenability*. This is the connection that, by holding between some information and a formula, A, makes the information eligible to be imported into the act of imagination whose explicit content is given by A (the term was used by Lewis [17] in the context of counterfactuals, and of course he took it from Goodman). The idea is that $[A]B$ will hold (at a world) when the explicitly imagined content, A, plus a *ceteris paribus* clause, say, C_A, dependent on A and cotenable with A (at that world), entails B. C_A is not an ordinary premise, or bunch of premises, but works rather like a catch-all *ceteris paribus* clause: it captures the background information we hold fixed relative to A, and which we can import into our imagined scenario.

What background is in fact imported is constrained by what is *relevant* with respect to the explicit content. Such relevance is difficult to pin down formally, but the intuitive insight is clear. Although I know that Amsterdam has more than 1,200 bridges, this is irrelevant with respect to my imagining that I fail my logic class: as my mental representation is not *about* that, I need not import the information about all those Amsterdam bridges. A kind of "aboutness" feature holds for our *ceteris paribus* operators, in the sense that $[A]B$ holds only when A and B share a sentential variable, so B is, in this sense, at least partly about what A is about: $[A]B$ is similar to a *relevant ceteris paribus* conditional, in the sense of relevance captured in relevant logics (see Priest [24], pp. 208–11). Showing this is not too difficult, given how weak the logic of our operators is.

And talking about weakness, here is the issue of the constraint hinted at above. So far I generically referred to what is imported in our exercises of *ceteris paribus* imagination as "information". But it is important to remark now that our cotenable information is not made only of *truths*,[10] because "what people do not change when they create a counterfactual alternative depends on their beliefs" (Byrne [5], p. 10), and believed falsities may get involved. As a consequence of this, (the counterpart in our framework of) what Lewis called Weak Centering should not hold in our semantics:

$$(\textit{Weak Centering}) \text{ If } w \in |A|, \text{ then } w \in f_A(w).$$

This says that if w realizes the explicit content A of an act of imagination, then it is one of the worlds picked out by the selection function for A. Even restricted to possible worlds, Weak Centering validates what we may call *imaginative modus ponens*:

$$A, [A]B \models_? B$$

[10] It is not even appropriate to call it *information*, if information is factive, as claimed by Floridi [12]. As the issue is controversial, we can, however, stick to a weak conception of semantic information as meaningful, well-formed data which need not represent reality correctly. Misinformation, in this sense, is a kind of information.

This says that if the explicit content of an act of imagination happens to really obtain, and it is represented in that act that B, then B also obtains. This is wrong because we can import false cotenable beliefs into our representation, as part of the relevant C_A. And this can make us imagine falsities even when A gets things right. For instance, I imagine that Obama works in Washington, $[A]$, but I mistakenly believe Washington to be in the state of Washington. I import the (relevant, cotenable) belief and I imagine Obama to work in the state of Washington, $[A]B$. A is true, but it does not follow that B is.

7 Indicative or Subjunctive?

The failure of imaginative *modus ponens*, and the doxastic component involved in the semantics for our operators, trigger another interesting issue which I will just briefly discuss, but leave open here: are these operators closer to indicative *ceteris paribus* conditionals than to subjunctive-counterfactual ones? While it has often been remarked that the logic of the two kinds of conditionals is very similar (e.g., they both fail Antecedent Strengthening, Transitivity, etc.), a key difference between indicatives and subjunctives is that what is cotenable with respect to indicatives is not made of facts, but of beliefs (see Bennett [1], p. 175–6). They are connected to subjective probabilities, or degrees of belief, so much so that according to some (including Bennett himself) one cannot even give a truth-conditional semantics for them.[11]

I suspect that our variably strict operators may behave in a way more similar to indicatives or to subjunctives depending on *how* the *ceteris paribus* worlds are selected. It might be, that is, that two different kinds of similarity or closeness are in play here. None of this has surfaced in this paper, precisely because I have introduced

[11] Even if indicative *ceteris paribus* conditionals lack truth values (which is controversial), one should not suspect that our *ceteris paribus* operators themselves lack genuine truth conditions, and thus that the whole prospect of a truth-conditional semantics as sketched here is flawed. Indicative conditionals, for authors like Bennett, lack truth values for they report or describe nothing, although they express something about the (conditional) belief arrangements of those who utter them. But a formula of the form $[A]B$ is exactly a report of the mental state of the relevant conceiving agent: it reports that the agent imagines that B (in a certain context) in the act of imagination whose explicit content is given by A; and such a report may be true or false. Anyway – and in answer to a point pressed by one anonymous Referee – the connections between the approach pursued in this paper, and the Ramsey test, which according to Bennett (and many others) explains the conditions for accepting or believing in *ceteris paribus* indicative conditionals, are presently not clear (to me, at least). Take beliefs as subjective probabilities. Then the Ramsey test connects the probability assigned to "if A, then B " to the conditional probability of B given A. The procedure, roughly put, is: to assess "If A, then B", one (1) takes the set of probabilities constituting one's present belief system, (2) adds probability = 1 for A, that is, full belief that A, (3) adjusts the rest of the belief system conservatively, making as few changes as possible to make room for A, and (4) sees whether the result gives a high probability for B. There is a similarity with the idea, described in this paper, of understanding "[A] B" as "Explicitly imagining that A, importing cotenable beliefs in the scenario, and getting B out of this". But I currently have no idea of how to fine-tune the connection.

no similarity metric of any kind for worlds. But I suspect that one may impose two different similarity structures, which would account for two different kinds of conceivability or imagination in the sense of Chalmers [6]: a *primary* conceivability where we imagine a certain scenario as a candidate for actuality, and which works in a way more similar to indicative *ceteris paribus* conditionals; and a *secondary* conceivability where we imagine a certain scenario as counterfactual, and which works closely to subjunctive conditionals in the sense of the relevant worlds similarity structure, although it differs (at least) in that Weak Centering is lacking. If such a development of the semantics presented above is feasible, it may nicely connect the framework to mainstream debates about conceivability and two-dimensional semantics. Whether the development *is* feasible will be, I hope, the topic of further work.

References

1. Bennett, J. (2003). *A philosophical guide to conditionals*. Oxford: Oxford University Press.
2. Berto, F. (2008). Modal meinongianism for fictional objects. *Metaphysica*, *9*, 205–218.
3. Berto, F. (2013). *The stanford encylopedia of philosophy*., Impossible Worlds Calif: Stanford University. http://plato.stanford.edu/entries/impossible-worlds/.
4. Brogaard, B., & Salerno, J. (2013). Remarks on counterpossibles. *Synthèse*, *190*, 639–660.
5. Byrne, R. (2005). *The rational imagination*. Cambridge: MIT Press.
6. Chalmers, D. (2002). Does conceivability entail possibility? In T. Gendler & J. Hawthorne (Eds.), *Conceivability and possibility* (pp. 145–199). Oxford: Oxford University Press.
7. Chellas, B. (1975). Basic conditional logic. *Journal of Philosophical Logic*, *4*, 133–153.
8. Costa-Leite, A. (2010). Logical properties of imagination. *Abstracta*, *6*, 103–116.
9. Delgrande, J. P. (1988). An approach to default reasoning based on a first-order conditional logic. *Artificial Intelligence*, *36*, 63–90.
10. Duží, M., Jespersen, B., & Materna, P. (2010). *Procedural semantics for hyperintensional logic*. Dordrecht: Springer.
11. Fagin, R., Halpern, J., Moses, Y., & Vardi, M. Y. (1995). *Reasoning about Knowledge*. Cambridge: MIT Press.
12. Floridi, L. (2005). Is information meaningful data? *Philosophy and Phenomenological Research*, *70*, 351–370.
13. Hintikka, J. (1962). *Knowledge and belief: an introduction to the logic of the two notions*. Ithaca, New York: Cornell University Press.
14. Jago, M. (2014). *The impossible: an essay on hyperintensionality*. Oxford: Oxford University Press.
15. Kahneman, D., Slovic, P., & Twersky, A. (1982). *Judgment under uncertainty*. Cambridge: Cambridge University Press.
16. King, J. (1996). Structured propositions and sentence structure. *Journal of Philosophical Logic*, *25*, 495–521.
17. Lewis, D. (1973). *Counterfactuals*. Oxford: Blackwell.
18. Lewis, D. (1978). Truth in fiction. *American philosophical quarterly*, *15*, 37–46.
19. Mares, E. (2004). *Relevant logic*. Cambridge: Cambridge University Press.
20. Meyer, R., & van der Hoek, W. (1995). *Epistemic logic for AI and computer science*. Cambridge: Cambridge University Press.
21. Niiniluoto, I. (1985). Imagination and fiction. *Journal of Semantics*, *4*, 209–222.
22. Nolan, D. (1997). Impossible worlds: a modest approach. *Notre Dame Journal of Formal Logic*, *38*, 535–572.

23. Priest, G. (2005). *Towards non-being*. Oxford: Oxford University Press.
24. Priest, G. (2008). *An introduction to non-classical logic* (2nd ed.). Cambridge: Cambridge University Press.
25. Rantala, V. (1982). Impossible world semantics and logical omniscience. *Acta Philosophica Fennica*, *35*, 106–115.
26. Ripley, D. (2012). Structures and circumstances: two ways to fine-grain propositions. *Synthèse*, *189*, 97–118.
27. Sainsbury, M. (2010). *Fiction and fictionalism*. London, New York: Routledge.
28. Wansing, H. (1990). A general possible worlds framework for reasoning about knowledge and belief. *Studia Logica*, *49*, 523–539.
29. Wansing, H. (2005). Connexive modal logic. In P.-H. Schmidt & W. Reynolds (Eds.), *Advances in modal logic* (Vol. 5, pp. 367–383). London: College Publications.
30. Wansing, H. (2014). *The stanford encyclopedia of philosophy*., Connexive Logic Calif: Stanford University. http://plato.stanford.edu/entries/logic-connexive/.
31. Williamson, T. (2007). *The philosophy of philosophy*. Oxford: Blackwell.

On the Preservation of Reliability

Bryson Brown

...all models are wrong, but some are useful.
(G E.P. Box and N.R. Draper, 1987)

Abstract "Mathematics may be compared to a mill of exquisite workmanship, which grinds you stuff of any degree of fineness; but, nevertheless, what you get out depends upon what you put in; and as the grandest mill in the world will not extract wheat-flour from peascod, so pages of formulae will not get a definite result out of loose date" (Thomas Huxley (1869) *Geological Reform*, Presidential Address to the Geological Society). Reasoning in science is a rich and complex phenomenon. On one hand, we find detailed, sophisticated and rigorous calculations. But on the other, we encounter a multiplicity of models and approximations whose status has been the subject of extensive debate (See [6] *How the Laws of Physics Lie* (Oxford, New York, Oxford University Press) and [7] *The Dappled World* (Cambridge, Cambridge University Press)). Detailed and demanding calculations give the appearance of mathematical rigour, and from a practical perspective, inferences and calculations based on successful models have proven to be reliable guides to our world, predicting the results of many measurements and suggesting interventions in the world that produce startling and impressive novel phenomena ranging from laser light to transistors to monoclonal antibodies and new types of sub-atomic particles. But the logical incompatibility of different models, each making different assumptions and approximations, together with the application of distinct, conflicting models in the course of deriving important results, raise serious questions about the nature and status of the both the premises and the conclusions of scientific reasoning.

Approximations and simplifications are often adopted without any demonstration that they are *faithful* to the principles whose implications they are intended to approx-

---

B. Brown (✉)
University of Lethbridge, Lethbridge, Canada
e-mail: brown@uleth.ca

H. Andreas and P. Verdée (eds.), *Logical Studies of Paraconsistent Reasoning in Science and Mathematics*, Trends in Logic 45,
DOI 10.1007/978-3-319-40220-8_4

imate.[1] Still, the successful application of the resulting models is widely adduced as support for the more precise principles. The logical concerns their practice raises are rarely addressed by the scientists in question, and the actual reasoning scientists do can be interpreted in different ways: in general, neither the kind of commitment they make to the formulae and sentences they write down nor the 'background' logic they draw on when making inferences are made explicit in their work.

A further interpretive challenge arises from the fact that reasoning in science is usually highly focused, seeking ways in which we might reason from certain assumptions to result(s) of particular interest. These may be results that have been observed and noted to be of particular interest already, such as atomic spectra or the precession of Mercury's perihelion, or results that are suggested by an interesting theoretical result, such as Bell's theorem. In general, efforts to systematically formulate the principles and to work out what can be inferred from them more generally come later, and only after a significant body of interesting results has emerged.[2]

Finally, the initial arguments are often sketchy and intuitive, even when some steps involve explicit calculations. For instance, Bohr's model of the hydrogen atom used calculations from classical physics to determine the energy differences between 'allowed' orbits constituting the possible states of the atom. But those explicit calculations provided no reason for restricting the states of the atom to those meeting the quantization of angular momentum rule and ignoring the rest of the continuum of orbits that could be modeled using the same classical principles.

When successful or even promising, such exploratory reasonings, with their approximations and shaky pro-tem assumptions, are often followed-up with more fully developed arguments, specifying, refining and sometimes generalizing the initial argument. Assumptions made at various points in the reasoning are gradually systematized, and more explicit calculations leading from the assumptions to conclusions of interest emerge.[3] Bohr managed to give an account of the hydrogen

[1]See [6], p. 119f., where Cartwright argues that the order in which a series of approximations each of which appears to be justified is to be made is chosen not on a principled basis (such as a theoretical argument for which ordering will produce more exact results), but instead on the basis of which order captures the observed Lamb shift in the excited state. Though in this case we can show one order produces a more precise approximation, in general this is not the case, and extra precision is not always required either. The subsequent discussion of the Lamb shift in the ground state provides an even more difficult case of a choice between approximations grounded in the result obtained rather than the mathematical precision of the approximations. Finally, a more general problem in the neighborhood is the lack of mathematical justification for the (brilliantly) successful calculational practice of re-normalization.

[2]Old quantum theory is a case in point: it developed via a series of strange proposals, beginning with the quantization of energy exchange between matter and the radiation field (Planck), and continuing with Bohr's hydrogen atom and its refinements and extensions (including ionized helium atoms), Ehrenfest's adiabatic principle (applying thermodynamics to connect stationary states of different physical systems), and more. See, for instance, Rechenberg, H. "Quanta and Quantum Mechanics," [19], pp. 143–248 in Laurie M. Brown, Abraham Pais, and Sir Brian Pippard, Eds., Twentieth Century Physics, Vol. 1, Bristol and New York: Institute of Physics Publishing and American Institute of Physics Press, for a helpful discussion of OQT.

[3]A famous cartoon by S. Harris in *American Scientist* provides an ironic illustration of this expectation—a scientist's long calculation on a blackboard is interrupted, at one point, by the

spectrum by combining restrictions on the allowed 'stationary states' with Planck's rule, $\Delta E = h\nu$. This success, with subsequent refinements and extensions (such as allowing for the finite ratio of the mass of the nucleus to that of the electron, applying the same sort of model but with a heavier nucleus with twice the elementary charge to capture the spectrum of ionized He atoms and invoking adiabatic transformations to link the quantization rules for various systems) gave rise to old quantum theory, leading to the subsequent emergence of quantum mechanics.

The logical ideal of a consequence relation determining the full set of commitments that follow from the adoption of certain premises is far removed from the give and take of scientific reasoning even once the assumptions of particular arguments are made clear and the reasoning from them to some important conclusions has been cast in precise and rigorous form. For example, classical mechanics gives rise to difficult puzzles including the inconsistency of a cosmological model combining infinite Euclidean space and a non-zero average density of the universe and Laraudogoitia's Zenonian supertask (and its temporal inverse) involving point particles.[4]

Philosophical studies exploring this difficult terrain have appeared in many places: Nancy Cartwright has argued that the approximations adopted in the course of reasoning from a background theory to models of some real systems are very often not *shown* to be reliable approximations of the theory they are based on; as a result, even when the models we actually apply to real systems are successful, we often cannot show that they capture the (approximate) *consequences* of the theory guiding the construction of our models of the systems. Instead, they bear a subtler, more independent relation to the theory or theories that they draw on.[5]

Further, models applied at different stages of a single calculation sometimes make incompatible assumptions, as in Bohr's 1913 model of the hydrogen atom.[6] Similarly, Mathias Frisch has argued that classical electrodynamics (at least as standardly applied) is outright inconsistent, by virtue of leaving out of account interactions that the theory requires.[7] Finally, Margaret Morrison has argued for a distinctive, independent role of models, separate from both theory and observation, in our understanding of science.[8]

(Footnote 3 continued)
words, "and then a miracle occurs," after which further calculations continue to a conclusion. Another comments drily, "I think you should be more explicit at this point".

[4]See Norton, "The Force of Newtonian Cosmology: Acceleration is Relative" *Philosophy of Science*, 62, [16], pp. 511–22, [17] "Classical Particle Dynamics, Indeterminism and a Supertask," *Synthese* 115: 259–265, and [9], 'Comments on Laraudogoitia's "Classical Particle Dynamics, Indeterminism and a Supertask", *British Journal for the Philosophy of Science*, 49: 123–133.

[5]See [6], p. 100ff, and [7], p. 179f.

[6]See Brown, B. and Priest, G., "Chunk and Permeate II: Bohr's hydrogen atom," *European Journal for the Philosophy of Science*, [3], doi:10.1007/s13194-014-0104-7.

[7][10] "Inconsistency in Classical Electrodynamics," *Philosophy of Science* 11/2004; 71:525–549. doi:10.1086/423627.

[8]Morgan, M. and Morrison, M. (eds.) [15], *Models as Mediators:Perspectives on Natural and Social Science*, Cambridge, Cambridge University Press.

All these issues raise questions about how we use theories and the concepts associated with them to reason about real systems and to report observations of them. This in turn raises doubts about the implications of both successful and failed empirical tests for the epistemic standing of a theory: when we are unsure about the inferences that link theories to empirically testable conclusions, it is hard to understand how and in what sense such theories come to be empirically confirmed or disconfirmed.

In general, the reasoning we do when we apply a theory to a system doesn't take the right form for a simple and straightforward approach: whatever logic we take to be truth-preserving, *truth* is not guaranteed to be preserved by the inferences we typically make when we make inferences from sentences in the language of some scientific theory to constraints on what we expect to observe. Thus from a strictly logical perspective, the observational disconfirmation of sentences we take to be testable 'implications' of a theory implies little if anything about the potential truth or falsity of the theory: in general (as Duhem and Quine[9] both pointed out) there are many *other* assumptions involved in such inferences. Similarly, when successful predictions don't, strictly speaking, *follow from* the theory, we are left in doubt about whether and how they count as successes *for the theory*.

1 Preservation of Reliability

"Preservationism" is the name of a logical school that began with work by Schotch and Jennings on their weakly aggregative logic called *forcing*. The forcing consequence relation preserves the *level* of premise sets (rather than the consistency of their consistent extensions), where the level of a set of sentences is an intuitive measure of the set's inconsistency in some underlying logic: a set of level 0 includes only theorems of the underlying logic; it can be consistently extended by any consistent collection of sentences of the language. A set of level 1 is a consistent set of sentences that cannot be consistently extended by any and every other consistent set of sentences, and a set of level n is a set that can be partitioned into, or *covered by* a minimum of n consistent sets. In general, preserving the level of a set of sentences requires us to weaken *aggregation* principles such as $\wedge$-I when inconsistency arises: for example, the level of $\{p, \neg p\}$ in classical logic is 2, but the classical level of $\{p, \neg p, p \wedge \neg p\}$ is ∞, the trivial level, since no consistent division of this set of sentences is possible. For n the level of a set of premises Γ, the strongest principle of aggregation that can be applied without trivializing Γ is 2/n+1, which forms the disjunction of pairwise conjunctions amongst any n sentences in Γ.[10] But any non-n colourable hypergraph will do as a template for complete aggregation, with the

[9]See [18], "Two Dogmas of Empiricism," *The Philosophical Review* 60: 20–43.

[10]See Brown and Apostoli, "A Solution to the Completeness Problem for Weakly Aggregative Modal Logic," *Journal of Symbolic Logic,* 60, 3, September 1995, 832–842 for the original completeness proof, and Brown and Schotch, "Logic and Aggregation" *Journal of Philosophical Logic* 28: 265–287 (June, 1999) [5] for a generalization in the context of hypergraph colourings.

points on each edge joined by conjunctions and the edges joined by disjunctions. The weakly aggregative approach to paraconsistency is not the only preservationist logic—for instance, preserving ambiguity measures instead of level leads to preservationist semantics for LP and FDE.[11]

In all these logics, the properties preserved by the consequence relation are *formal* properties of the premises we reason with. In this paper, I propose a preservationist perspective on reasoning in science, but the property I suggest scientific reasoning aims to preserve is *reliability*. Since reliability is not a formal property like level, consistency, or ambiguity measures, there can be no formal account of the inferences that preserve it. But I think that considering some concrete examples provides interesting insight into what this pragmatic, preservationist perspective can contribute to our understanding of science. In particular, the reliabilist perspective points towards a modest form of scientific realism focused on the present state of science, rather than on what we conceive as the ultimate aim of science.

To better understand reasoning in science we need a broader view of how various concepts and the inferences that we make based on them contribute to the models scientists apply when reasoning about various phenomena. Our focus here will be less on theories, whether conceived syntactically as sets of sentences closed under a consequence relation or as sets of models, and more on specific examples of scientific reasoning and the way in which various concepts are *deployed* in the course of reasoning. In general, concepts can be used both to report the outcomes of various observations and in the course of scientific reasoning, where scientists rely on inferential connections between them, sometimes in the form of general laws ($f = ma$, for example) and sometimes in less rigorous patterns of inference (as when noticeable differences in appearance between two groups of organisms are taken as evidence that they represent distinct species).

The inferential connections involved are often too complex for a rigorous application to most, or even all real systems.[12] However, in many cases simplified accounts of real systems in terms of these concepts can be applied to build *reliable* models of such systems, that is, models from which we can reliably infer sentences that *satisfactorily* describe observed properties and relations of the systems.[13] Furthermore, the standards for what counts as a satisfactory description shift as improved observations reveal details and relations that had not been noted or seen as significant

---

[11]See "Ambiguity Games and Preserving Ambiguity Measures," in *On Preserving: Essays on Preservationism and Paraconsistent Logic*, Schotch, Brown and Jennings, eds., University of Toronto Press (2009).

[12]In fact, even where rigorous general accounts are available, they typically emerge from careful reflection on scientific practices applicable to particular cases which were quite successful long before those rigorous accounts emerged. Consider as an example the development of quantum mechanics leading up to von Neumann's formal account.

[13]We will say a description is satisfactory if what it says about the system is approximately true in the innocent sense of agreeing, within contextually determined limits, with reports of various observations of the system, expressed in terms of the same concepts.

before.[14] In other cases, while natural systems remain too complex for precise and reliable application of models built with the help of such concepts, scientists are able to produce carefully designed and built systems that do display behaviour that fits with the principles derived from some such model.[15]

When these kinds of successes occur, scientists take the concepts that underlie the models in question *seriously* as part of our efforts to describe and reason about the systems being modeled. While the commitments scientists take towards these concepts and the models built with their help range from very pragmatic and local to strongly realistic commitments with very broad scope, scientists across this range of attitudes often adopt these concepts and inference rules as tools worth refining, applying them to reasoning about other systems and testing them in more detailed and precise ways. Models applying such concepts are also relied on in the production of new kinds of real systems, aimed at producing effects of interest that the model suggests such systems will give rise to.[16] Reliability is clearly a central value here: concepts that can be relied on to

Even when calculations within a model approximate in ways that limit the application of a theory's conceptual apparatus within the model, and we cannot in principle show that a conclusion that emerges will in fact approximate the results of a full application of the theory's principles,[17] successful efforts to reliably *produce* patterns of observable phenomena previously identified as 'theoretical' possibilities by inference from the model is widely seen as providing significant support (of some kind) for the theory.[18]

As I see it, a theory which provides conceptual resources for the construction of such empirically successful models deserves some kind of credit, but this credit does not justify the claim that the theory constitutes a set of sentences likely to be *true* of the world. Instead, it justifies regarding the theory as a valuable and reliable *cognitive*

---

[14] Consider the impact of Brahe's observations on astronomy and Kepler's efforts to improve on Copernicus' model for planetary motion; the shift to ellipses governed by Kepler's laws allowed for a much better fit with the observations. Another example is Einstein's focus, as he was working towards his theory of General Relativity, on capturing the precession of Mercury's perihelion.

[15] See Cartwright on 'phenomena' in [6], p. 100ff.

[16] Think here of efforts to produce 'effects' predicted by simple theoretical models—masers and lasers, transistors and many other basic components of modern electronic and optical technology are concrete demonstrations of phenomena that theoretical results had identified as potentially realizable.

[17] For example, we may sum the first few terms of a series whose terms quickly become quite small, and accept the result as an 'approximation' to the sum of the series as a whole without a formal proof that the series actually converges to a limit. Successful application of such results are often counted as a positive result for the theory in question, even though it's possible that they are not in fact good approximations to what the theory would predict if a more rigorous calculation were performed.

[18] What I have in mind here is related to Cartwright's account of *phenomenal laws* (*How the laws of physics lie*), which describe reliable regularities that hold of phenomena we either find or learn to create. Such laws are not strictly *derived* from the theory's principles—instead, they invoke concepts drawn from the theory, reasoning with them in ways that are not logically rigorous, but which *might* (at least approximately) capture results that could, in principle, be rigorously derived; one might say they are *inspired* by the theory rather than derived from it.

resource. The concepts involved in such cases have an established track record of leading scientists to infer striking patterns of phenomena that are either observed to occur in nature or can be successfully produced in some way or other. They are often reliable 'maps' of what we can expect to observe in various circumstances.

In her work Cartwright has emphasized the importance of the practical challenges scientists face, as they learn to reliably produce and observe certain phenomena. My point here is that it is similarly difficult to extract useful models from a theory's principles and to discern how to apply new concepts to real systems, reasoning with them in a way that leads to empirically significant conclusions, including predictions of various phenomena.[19] Both require ingenuity, insight into the theory and into the apparatus or natural system involved, together with clever choices allowing scientists to simplify and extract a useful tool (a model or a system of objects and instruments) from something too complex to be completely understood (a theory or some part of the natural world). Thus at both ends of a successful encounter between our representations and a system in the natural world we find creative insight, trial and error and rich interaction between scientists, representations and features of the world those representations can be usefully applied to.

I also want to emphasize here that there are, logically speaking, *messy* cases that deserve careful examination. Bohr's model of the hydrogen atom boldly ignored the implications of classical electrodynamics for his postulated orbiting electrons, while assuming that classical electrodynamics can be relied on when making observations of the radiation emitted by a sample of excited hydrogen gas: Bohr's inference to empirical claims about the hydrogen spectrum begins with a model of the atom and its states that rejects electrodynamics, allowing the calculation of a 'frequency' associated with a transition between two 'stationary states' based on Planck's equation, $E = h\nu$. But what is this frequency a frequency *of*? Bohr depended on classical electrodynamics to guide him here, enabling him to connect the frequencies derived from his quantum model with empirically testable claims about the radiation field around a sample of excited hydrogen atoms.[20] Successful applications came quickly- not just an account of the known hydrogen spectrum, but a prediction of previously unknown lines, a successful refinement taking account of the finite mass of the nucleus and achieving a better fit with spectral data, an account of new lines in the solar spectrum as due to singly-ionized helium atoms, and an account of line splitting in terms of highly elliptical, relativistic orbits.[21] But these further models of the old quantum theory also made predictions about spectra by setting aside electrodynamics when dealing with atoms, their states and transitions between states, while those predictions continued to be interpreted as predictions about classical electromagnetic radiation,

---

[19] Consider Bell's work on non-locality in QM (see [1], "On the Einstein Podolsky Rosen Paradox," Physics 1 (3), 195–200): his discovery that the statistics of such QM observations would differ in an experimentally testable way from those of a hidden-variable theory demonstrated the possibility of settling experimentally what had, up to that point, been widely thought of as a metaphysical issue.

[20] Brown and Priest, "Chunk and Permeate II: Bohr's Hydrogen Atom," *European Journal for Philosophy of Science*, Jan 2015 [3]. http://link.springer.com/article/10.1007/s13194-014-0104-7.

[21] Ibid.

which provided the only available models of how light interacts with instruments to give rise to observable spectra.

The reasoning employed by Bohr was selective, targeted and *speculative* in spirit. It identified interesting conclusions that could be drawn from assumptions that were applied in such a way as to avoid known difficulties. The elixir of Planck's equation provided a bridge between Bohr's quantized, semi-classical model of the atom and claims about the radiation emitted by excited hydrogen atoms that we already understand how to test; Einstein famously described it as the "highest form of musicality in the sphere of thought."[22]

The reasoning involved here isn't captured by the standard logical model of a consequence relation, in which the commitments that premises carry with them are expressed in terms of the *closure* of the premises under the consequence relation, and any course of reasoning that combines the premises employed and conclusions drawn logically from them in any way is endorsed. Instead, Bohr applied his premises *strategically*, at specific points in the course of the inferences leading from descriptions of the atoms to descriptions of the radiation they emit. The premises, taken altogether, were contradictory: the orbiting electron in a stationary state fails to radiate, but the radiation emitted in a transition from a higher energy state to a lower, which was essential to actual empirical testing of the model, was still described in terms of classical electrodynamics. But these conflicting principles were applied at different points in the course of Bohr's reasoning: the calculations invoke both, but never together. The success of Bohr's model demonstrated that his model was a *reliable inferential tool* when applied to the hydrogen atom *in accord with the restrictions* Bohr had imposed. In the end, the model produced important, reliable results where none had been before[23]—a success that led to more and increasingly systematic exploration of what could be achieved with the combination of classical and quantum principles that characterized old quantum theory.

2 Other Examples:

Although they don't invoke incompatible basic laws, complex models such as regional climate models (RCMs) display similar patterns of reasoning. In RCMs[24] high resolution regional models are 'nested' within relatively low-resolution global climate models (GCMs) in order to explore the potential impacts of global climate change on local climate patterns, including precipitation, temperature, maximum

[22] Schilpp, Paul Arthur, editor. *Albert Einstein: Philosopher-Scientist*, pp. 19, 21, Open Court, La Salle, Illinois, [1949; 1951] 1969, 1970. ISBN 0-87548-286-4.

[23] In fact some had thought no such results could be expected, since, if these lines characterized 'resonant' frequencies of a tiny, complex system, the mathematics of determining the structure of the system responsible for them (as in calculating the shape of a bell from its sound) seemed beyond solution.

[24] See http://www.ouranos.ca/en/scientific-program/climate-science/climate-simulations/ and http://www2.mmm.ucar.edu/wrf/users/tutorial/200807/WRFNesting.pdf.

flood levels etc. In model runs, the 'coupling' of the two takes place across a geographical boundary where the conditions projected by the RCM are reconciled with the conditions projected by the GCM, which 'drives' the regional model at its lateral boundaries. Just as Planck's law provided a bridge between Bohr's model of the hydrogen atom and a (strictly incompatible) CED model of the radiation field surrounding a sample of excited hydrogen atoms, imposing the results of the GCM as boundary conditions in the RCM allows the GCM to affect the development of the RCM over time.

Once again reasoning here does not proceed on the standard logical model of a consequence relation- that is, it does not draw on a single collection of premises and endorse every inference those premises would license as correct. In one-way nesting, results for the RCM are arrived at by applying the GCM to calculate conditions (temperature, pressure, etc.) at the *lateral boundary* of the nested RCM. The figures from each cell of the GCM are imposed as boundary conditions on the corresponding cells of the RCM (with an integer number, usually odd, of RCM cells aligned with each GCM cell) through a 'translation' that interpolates steps in values across the corresponding GCM grid squares and imposes the finer-scale topography of the RCM. The RCM is then run to produce results for its area of coverage over a time step, after which the RCM's boundary conditions are again adjusted to reflect the results derived from the GCM. In two-way nesting, the results of some number of time-steps for the RCM feed back into the boundary GCM cells, affecting the subsequent results in the GCM.

The phenomenon I want to focus on is the limited interaction between the GCM and the RCM. Parameter values are passed, according to systematic rules, from one to the other (from the GCM to the RCM for one-way nesting, and in both directions for two-way nesting). The equations used in each don't cross over; instead, they are applied to the parameters of a given state to calculate a new state at each time step in each model. To capture this kind of reasoning, we need models of inferential systems which allow the separation of contexts together with the passing of some results from one context to another. One such model is 'chunk and permeate' (C&P)[25]: in this model, separate 'chunks' contain different (sometimes inconsistent) premises, and reasoning processes allow some results to 'permeate' from one cell to another. A cycle of reasoning begins by closing each chunk under its consequence relation, following which specified kinds of sentences in each chunk permeate to certain other chunks; the 'consequences' of a C&P structure are the sentences found in a particular chunk (called the *designated chunk*) in the limit of these cycles.

More importantly for our purposes, justifying such a reasoning practice requires an account of how, why, when and *in what sense* it can be said to work. The standard semantic model of a consequence relation relies on guaranteed truth preservation: if certain premises are true, all (and only) their consequences are guaranteed to be true as well. This 'guarantee' is achieved straightforwardly: by definition, $\Gamma \models \alpha$ iff

[25] See Brown and Priest, "Chunk and Permeate, A Paraconsistent Inference Strategy, Part I: The Infinitesimal Calculus," *Journal of Philosophical Logic*, 33, 379–388, 2004 [2], and Brown and Priest [3], op. cit..

α is satisfied in every model of Γ; syntactically, $\Gamma \vdash \alpha$ if α is consistent with every consistent extension of Γ. This also implies that $\models$ and $\vdash$ are *closure relations*—that is, $\{\gamma : \Gamma \models \gamma\} \models \alpha$ iff $\Gamma \models \alpha$ and $\{\gamma : \Gamma \vdash \gamma\} \vdash \alpha$ iff $\Gamma \vdash \alpha$.

But preservationism provides a broader approach to reasoning. Unlike other examples of preservationist systems, the preservationist suggestion here extends beyond the formal accounts of a consequence relation: I propose that we should regard the kind of reasoning I've been discussing here as aimed at the preservation of *reliability*. That is, the inferences countenanced in successful models of these kinds are inferences that are found to produce reliable conclusions about the systems we apply them to. From this perspective 'truth' is a limit concept, corresponding to *reliability with no holds barred*: a worthy goal, but its 'all of nothing' nature makes it a poor measure of progress.

In this light, since chunk and permeate gathers its conclusions in the *designated* chunk, to evaluate an instance of chunk and permeate reasoning we need to consider the reliability of the sentences appearing there. In the case of Bohr's hydrogen atom, the designated chunk is the classical chunk, where spectra are predicted and their observation can be reported. So on this account, Bohr's model was a success because it generated a collection of empirically reliable results, including agreement with established results about the hydrogen spectrum as well as predictions that were subsequently found to be empirically reliable.[26] Extensions and refinements of the model and other related models that produced still more reliable results across a wider scope of systems emerged subsequently, as old quantum theory developed. As a result, the Bohr model together with an increasingly systematized collection of models using related methods came to be adopted as tools producing empirically reliable inferences where none had been available before. Absent the initial reliable results, Bohr's model and its implications for the hydrogen spectrum would surely have been still born; one suspects it would never have escaped Bohr's working notebooks.

The empirical reliability of this kind of reasoning cannot be a matter of *truth preservation*: as a whole, the collection of premises used in the course of Bohr's reasoning was never a candidate for truth. Further, Bohr's model didn't match all the features of the empirical spectra (for example, that the observed spectral lines are broadened by the short half-lives of highly energetic states[27])—it could never have been a candidate for the 'whole truth' about the hydrogen atom. Bohr's success turned on the unprecedented and quite detailed partial agreement between accepted spectral observations and the consequences derived from his model. The results of reasoning with the model in the way prescribed by Bohr agreed with established regularities about certain spectra as well as with subsequent observations of spectral lines that hadn't been previously observed. These observations in turn were

[26]This went beyond spectral data to include a satisfactory estimate of the typical size of a hydrogen atom, and an explanation of why lines due very high-energy states were missing from known spectral studies, due to the orbital radius of such states and their consequent instability/absence at normal pressures.

[27]On this topic see, for example, [11]. "Observation of Inhibited Spontaneous Emission". *Physical Review Letters* **55** (1): 67–70, in which inhibition of emission is used to narrow spectral line-widths.

considered reliable based on an established record of agreement (and resolution of disagreements) on a range of observations of spectra. These in their turn relied on an established background of reliable, independently replicated results including methods for producing gratings and other optical equipment, along with lab procedures documented and transmitted in the training of new spectroscopists. From the point of view proposed here, the reliability of such empirical observational practices and the practices involved in constructing and using reliable instruments is neither more nor less fundamental (and neither more nor less to be taken for granted) than the reliability of the inferences scientists make with the help of theories and models.

The example of regional climate models is subtler still, since the aim of these models is not straightforwardly predictive; instead, they aim to provide policy guidance for government, agencies, corporation and individuals by identifying *possible* changes in maximum precipitation, spring flood levels and other important climatological variables. Reliability here is hard to determine directly, since we only get one 'real' trial, and it is yet to be realized. But indirect tests are possible: we can reliably fill deliberately imposed gaps in past temperature data, using both regional and global models; we can try to model the periodicity of past cycles of glacial and interglacial periods based on models of variations in distribution of insolation due to Milankovitch cycles combined with models of various feedbacks; we can retrospectively impose unpredictable boundary conditions including ENSO and volcanic eruptions to see whether their addition to our models produces a better fit to the actual record, and we can improve data bases by developing better interpolation techniques for thinly covered regions.[28] The results suggest that the guidance provided by these models is better than (i.e. *more reliable than*) guessing as a guide to risk management and public policy.

Many important inferences in science are not easily understood as preserving the truth of some accepted set of premises, i.e. as grounded in the fact that *if* the specified premises are true, the conclusion must also be true. But if we focus on reliability as a pragmatic property preserved by successful scientific inferences we see immediately that it also applies, with limitations, to models long since relegated to the 'junk pile' of serious science, including earth-centred astronomy and phlogiston chemistry. I suggest that these models have largely[29] fallen out of use, not because they aren't reliable in certain conditions and within certain limits, but because we have more generally reliable models whose use does not make significantly greater demands on our ability to reason with them or compare them with the results of observation. By contrast, we still confidently use models based on classical physics in many scientific contexts, because they are both sufficiently reliable and simpler to apply than models based on special or general relativity (such as in planning space probe missions) or on quantum mechanics (for designing magnetic resonance imaging systems).

[28] [8], Coverage bias in the HadCRUT4 temperature series and its impact on recent temperature trends. Q.J.R. Meteorol. Soc. doi:10.1002/qj.2297.

[29] Of course for navigational purposes, earth-centered astronomy is still a convenient tool.

3 Preservation of Reliability, Scientific Revolutions and Scientific Realism

Debates over scientific realism and scientific progress often focus on the question of what is *preserved* over time in the development of science. Natural languages provide persistent and apparently stable ways of describing our world, and we can apply them to express much of what we've learned about the natural world in the course of scientific inquiry. In contrast to this stability, history shows that scientific theories are supplanted, over time, by distinct successor theories. This instability has been a crucial premise in some familiar arguments against scientific realism, while the apparent stability of the language we learn 'at mother's knee' has been a strong point in favour of a kind of common-sense realism about the world view Sellars called the "manifest image".[30] The question of what is preserved and progressively extended as science 'advances' seems easier to answer in natural language, while *scientific realism* in all its forms has included a preference for the account of the world expressed in the theoretical language(s) of science over one expressed in natural language, even a natural language that has been carefully refined to make it more philosophically clear and coherent.

Identifying some kind of epistemic accomplishment or success that is preserved across changes of theory and that has increased over the course of the history of science would provide a basis for the intuitively obvious claim that science is a progressive enterprise. As a failed example, consider the logical empiricist distinction between the observation language and theoretical languages: whatever its failings, it explicitly allowed for the preservation and extension of observational successes across theoretical changes. But in the latter half of the 20th century, the inseparability of the language of observation from the language of scientific theory came to be widely recognized. The actual observation reports of scientists could not be 'cashed in' either in terms of a phenomenalist language or in terms of a philosophically refined version of the common-sense language of everyday observations.

This new perspective raised difficult questions about what sort of 'content' is preserved across different scientific languages and scientific revolutions.[31] A new form of realism, inspired in part by new ideas about names and natural kind terms, made the preservation of successful *reference* for theoretical terms central to its account of scientific progress. But counter-arguments emerged quickly: many theoretical entities belonging to 'mature' sciences, such as the 'luminiferous ether,' were subsequently dropped from our scientific ontology. Thus even mature sciences don't provide a reliable partial inventory of *what there is* that is preserved over time: transformations

---

30 Sellars, W., "Philosophy and the Scientific Image of Man," in *Frontiers of Science and Philosophy*, Robert Colodny (ed.) Pittsburgh, PA: University of Pittsburgh Press, (1962), 35–78; reprinted in *Science, Perception and Reality*, London: Routledge and Kegan Paul, New York and The Humanities Press [20], 1–40.

31 (see T.S. Kuhn *The Structure of Scientific Revolutions*, 2nd ed., Chicago, University of Chicago Press, 1972).

of our theories' conceptual structure can drop terms that were once taken to refer to fundamental (even *indispensable*) entities.[32]

Bas van Fraassen's empiricism[33] presents a view of science as aimed, not at truth or at an inventory of existing items, but more subtly at the pursuit of 'empirical adequacy', something that science could come closer to achieving by replacing theories shown to be in conflict with settled observations with theories that are compatible with them. Since empirical adequacy can be achieved very simply by imposing no constraints on the models of a theory at all, van Fraassen's account also acknowledges (though it does not emphasize) the value of *empirical strength*, whose preservation would require retaining *empirically successful predictions* of a new theory's predecessors. And it's clear that such successes have (at least often) been retained.[34] For example, the replacement of classical gravity by general relativity preserved classical gravity's success in accounting for planetary orbits (and many other phenomena) while improving on classical gravity accounting for the 'anomalous' precession of Mercury's perihelion and the displacement of the angular positions of stars as observed by Eddington during the solar eclipse of 1919.

But here too, preservation and progress are hard to establish: like truth, empirical adequacy is an ideal goal, demanding a perfect accord between the empirical substructure of some model of the theory with the observable structures of the world. Van Fraassen's constructive empiricism departs from realism only by restricting the required accord between theory and world to what *could be observed* by human beings at the right time and place.[35] Van Fraassen explains empirical adequacy in terms of what he calls the *empirical substructures* of the models of a theory. These are the substructures that, according to the theory's model of human beings and their sensory capacities, can be directly detected by a human being at the right place and time, using only her senses. Thus reports constraining what models of a theory are such that their empirical substructure is compatible with our observations can be correctly made as an immediate cognitive response to being exposed to certain kinds of systems as described by the theory, by human beings trained to make observations in the language of the theory. A theory is *empirically adequate* if at least one of the models of a theory satisfies all the observation reports that could made by (properly trained) human beings if they were present, where those reports are limited to what

[32]Laudan, Larry. "A Confutation of Convergent Realism", *Philosophy of Science*, Vol. 48, No. 1, (Mar. [14]): 19–49.

[33]*The Scientific Image,* Oxford, Clarendon Press, 1980.

[34]Though examples of "Kuhn loss," *The Structure of Scientific Revolutions*, Chicago: University of Chicago Press ([13], 2nd edition, with postscript) 99–100), i.e. the surrender of what seemed to be successful explanations in the transition to new theories, have been proposed including the purported explanation of the similarities of different metals (shininess, ductility etc.) on the basis of their containing phlogiston, I do not pursue this issue further here.

[35][2], "The Pragmatics of Empirical Adequacy," *The Australasian Journal of Philosophy*, 82, 2, 242–264.

their sensory capacities, as described by the theory's models of human beings, allow them to detect.[36]

We have here what looks to be good evidence for progress in terms of empirical adequacy and strength with respect to currently accepted observations: scientists generally favour theories that have survived empirical testing to date, along with theories that *constrain* empirical results more strongly without conflicting with accepted observations. But the way we go about using theories in practice is open to much looser interpretation: in practice, there is a back-and-forth between efforts to produce acceptable observations and the models we expect those observations to accord with. Cartwright[37] emphasizes that while models draw on theoretical concepts/ principles, they make substantial compromises along the way: often we cannot prove, in principle, that inferences made from models really are good approximations to what strictly follows from the theory. The implications for pro-tem judgements of the empirical adequacy of a theory in relation to specific phenomena are difficult to reconcile with confidence about steady progress towards empirical adequacy. We may well conclude that some observations refute a theory's empirical adequacy, or that they are compatible with it, only to find later, as our observations and calculations are refined, that this is not the case. On the other hand, Cartwright's phenomenal laws[38] hold in highly controlled/ specific conditions which often involve complex observational practices. They may reliably characterize the phenomena they apply to, but they are not derived from theories, and the observational practices that produce the phenomena are not straightforward 'recipes' either. Consequently, for Cartwright, theories get little credit when it comes to truth claims—however, their value as starting points and conceptual 'toolkits' remains.

Nevertheless, from Cartwright's anti-realist perspective successful theories are *repositories* of concepts and inference patterns that scientists rely on to guide the construction of reliable models, and to help us to discover and reliably produce a range of phenomena.

The point I want to make here is that there is a subtler kind of realism, focused in the here and now rather than on some conception of the ultimate aim of science. Science as it stands today, with its rich variety of theoretical apparatus and concepts, its complex observational instruments and methods, its regular phenomena and models that account for them, constitutes a better, more powerful and unified

[36] This is not an easy account to follow through on. It is difficult to see how we can determine what observations humans can make in the language of a theory when we don't know how to describe ourselves in the language of the theory without the help of a very substantial body of observations.

Setting aside this worry about epistemic circularity in hopes that a more pragmatic approach could dissolve the problem, we still need to assume that a pragmatic approach to observation would not entirely undermine the assumption of a privileged epistemic status for observations made using unaided human senses. And at that point we would still need work out how to categorize the many things that are, intuitively, humanly observable (that we believe would trigger a distinct sensory response in a human being to whom they were *present*), but which occur in situations such that a properly trained human being who was present would not survive long enough to actually recognize and interpret her sensory response. See the discussion in [2], op. cit.

[37] [7], op. cit.

[38] [6], op. cit.

description of the world than an account framed in the concepts provided by our natural human languages. It is more precise and more powerful; it has given rise to methods for producing startlingly unintuitive and practically valuable phenomena, from antibiotics and microchips to aircraft, generators and electrical grids. It has predicted strange and wonderful things, such as the cosmic microwave background radiation. Our understanding of the world would be deeply impoverished by a philosophical attempt to push our cognitive commitments back into the straightjacket of common sense, or even that of Sellars' refined 'manifest image.'[39]

4 Conclusion

We do not know in advance what level of local, detailed reliability and breadth of applicability can be achieved. The world need not have been as predictable, as reliably and systematically *describable* or as effectively *manipulable* as it has turned out to be (of course for all we knew, the task could also have turned out to be easier). The emergence of modern science is, I suggest, best understood in terms of our having stumbled across a 'sweet spot' in terms of both our concepts and the development and application of instruments in observation, and, more recently, in calculation. In retrospect, the path to contemporary science looks fraught: systems such as Euclidean geometry and classical mechanics were fundamentally misleading, but still served as helpful—even indispensable—stepping stones: despite being false, they were and remain reliable as guides to reasoning about our world in many circumstances. Better still, many of their features were retained in the later theories that emerged to replace them, even though there seems to be no basis for *ruling out* the possibility of a world in which such local, limited theoretical successes don't occur, or in which they occur but don't provide a helpful 'stepping stone' towards better theories. In addition, empirical measures have emerged that reliably identify conditions under which, and the degree to which particular models, conceptual apparatus and judgments can be relied on. Physical parameters we can use to do this include temperature, the strength of gravitational fields as well as scales of size and relative velocities, but psychologists have studied examples touching directly on human cognitive behaviour, revealing, for instance, reliable patterns of perceptual and cognitive errors we tend to make as well as standards and methods we can apply to avoid them (though, sadly, we often fail to apply them).[40, 41]

An underlying philosophical uncertainty persists: setting aside the known limitations, we cannot rule out the presence of other, unknown factors which undermine the reliability of a thus-far reliable inference or observation in a particular case, even when it respects the known limits of such inferences. But this is just a generalization

[39]Sellars, op. cit.

[40]See [12], *Thinking, Fast and Slow* (Doubleday) for a rich overview of some of this work.

[41]For climate simulations and nesting, http://www.ouranos.ca/en/scientific-program/climate-science/climate simulations/, http://www2.mmm.ucar.edu/wrf/users/tutorial/200807/WRFNesting.pdf.

of Hume's worry about induction. From a pragmatist perspective, which recognizes the indispensability of induction in our confidence regarding both observation and reasoning, it justifies a fallibilist modesty about our judgments, but not despair about our ability to *justify* them.

References

1. Bell, J. (1964). On the Einstein Podolsky Rosen paradox. *Physics*, *1*(3), 195–200.
2. Brown, B. (2004). The pragmatics of empirical adequacy. *The Australasian Journal of Philosophy*, *82*(2), 242–264.
3. Bryson, B., & Graham, P. (2015). Chunk and permeate II: Bohr's hydrogen atom. *European Journal for the Philosophy of Science*, *2015*. doi:10.1007/s13194-014-0104-7.
4. Brown, B., & Apostoli, P. (1995). A solution to the completeness problem for weakly aggregative modal logic. *Journal of Symbolic Logic*, *60*(3), 832–842.
5. Brown, B., & Schotch, P. (1999). Logic and aggregation. *Journal of Philosophical Logic*, *28*, 265–287.
6. Cartwright, N. (1983). *How the Laws of Physics Lie*. Oxford, New York: Oxford University Press.
7. Cartwright, N. (1999). *The dappled world*. Cambridge: Cambridge University Press.
8. Cowtan, K., & Way, R. G. (2014). *Coverage bias in the HadCRUT4 temperature series and its impact on recent temperature trends*. United Kingdom: Royal Meteorological Society. doi:10.1002/qj.2297.
9. Earman, J., & Norton, J. D. (1998). Comments on Laraudogoitia's "classical particle dynamics, indeterminism and a supertask". *British Journal for the Philosophy of Science*, *49*, 123–133.
10. Frisch, M. (2004). Inconsistency in classical electrodynamics. *Philosophy of Science*, *71*, 525–549. doi:10.1086/423627.
11. Gabrielse, G., & Dehmelt, H. (1985). Observation of inhibited spontaneous emission. *Physical Review Letters*, *55*(1), 67–70.
12. Kahneman, D. (2011). *Thinking, fast and slow*. New York: Farrar, Straus and Giroux.
13. Thomas, K. (1970). *The structure of scientific revolutions* (2nd ed.). Chicago: University of Chicago Press.
14. Laudan, L. (1981). A confutation of convergent realism. *Philosophy of Science*, *48*(1), 19–49.
15. Morgan, M. S., & Morrison, M. (1999). *Models as mediators: perspectives on natural and social science*. Cambridge: Cambridge University Press.
16. Norton, J. D. (1995). The force of newtonian cosmology: acceleration is relative. *Philosophy of Science*, *62*(1995), 511–522.
17. Pérez Laraudogoitia, J. (1998). Classical particle dynamics, indeterminism and a supertask. *Synthese*, *115*, 259–265.
18. Quine, W. V. O. (1951). Two dogmas of empiricism. *The Philosophical Review*, *60*, 20–43.
19. Rechenberg, H. (1995). Twentieth century physics. In M. Laurie, A. P. Brown, & B. P. Sir (Eds.), *Quanta and quantum mechanics* (pp. 143–248). Bristol and New York: Institute of Physics Publishing and American Institute of Physics Press.
20. Sellars, W. (1962) Philosophy and the scientific image of man. In: Colodny (ed.), *Frontiers of science and philosophy* (pp. 35-78), Pittsburgh: University of Pittsburgh Press. (Reprinted from Sellars, *Science, perception and reality*, (pp. 1-40). London and New York: Routledge and Kegan Paul (1963)
21. van Fraassen, B. (1980). *The scientific image*. Oxford: Clarendon Press.

Prospects for Triviality

Luis Estrada-González

For Christian Edward Mortensen, long-distance mentor, in his 70th birthday.

Abstract In this paper I argue, *contra* Mortensen, that there is a case, namely that of a degenerate topos, an extremely simple mathematical universe in which everything is true, in which no mathematical "catastrophe" is implied by mathematical triviality. I will show that either one of the premises of Dunn's trivialization result for real number theory –on which Mortensen mounts his case– cannot obtain (from a point of view "external" to the universe) and thus the argument is unsound, or that it obtains in calculations "internal" to such trivial universe and the theory associated, yet the calculations are possible and meaningful albeit extremely simple.

Keywords Triviality · Atomic triviality · Real number theory · Degenerate categories · Internal logic

---

This paper has been written under the support from the PAPIIT project IA401015 "Tras las consecuencias. Una visión universalista de la lógica (I)", as well as from the CONACyT project CCB 2011 166502 "Aspectos filosóficos de la modalidad". I thank Charlie Donahue and Chris Mortensen for useful comments on previous versions of this paper, as well as to the referees for saving me from at least a couple of embarrassing mistakes. Diagrams were drawn using Paul Taylor's diagrams package v. 3.94.

L. Estrada-González (✉)
Instituto de Investigaciones Filosóficas, Universidad Nacional Autónoma de México, Circuito Maestro Mario de la Cueva s/n Ciudad Universitaria, C.P. 04510 Coyoacan, Mexico, D.F., Mexico
e-mail: loisayaxsegrob@gmail.com

H. Andreas and P. Verdée (eds.), *Logical Studies of Paraconsistent Reasoning in Science and Mathematics*, Trends in Logic 45,
DOI 10.1007/978-3-319-40220-8_5

1 Introduction. The Anathema Against Triviality in Mathematics

One of the main outcomes of the growth and settlement of studies on paraconsistency is inconsistent mathematics, which is now a well-established field of mathematical research. One a more methodological vein, the studies on paraconsistency propitiated an adequate degree of intellectual freedom for honestly questioning the necessity of any principle of logic, no matter how venerated it had been at some point. Some paraconsistentists and their interlocutors (e.g. Priest's [14, 16]), Mortensen's [9, 12], and Kabay's [4]) have lately given a certain amount of airplay to trivialism, the idea that everything is true. However, and although nothing is completely uncontentious, the almost unanimous view is that trivialism is wrong.

In this short paper I discuss trivialism restricted to mathematics, that is, whether there could be place for trivial theories (those where everything is true) and trivial objects (those with every property) in mathematics. Again, the almost unanimous view is that a trivial mathematical theory is the "worst sort of expansion [a mathematical theory might have]" [1, p. 323] and that triviality in mathematics is "absolutely unacceptable" [19, p. 612], an "Armageddon" [2, p. 1], to take just three random verdicts on triviality in mathematics from among friends of inconsistent mathematics. Even Mortensen, who is overall more sympathetic to trivialism in general, considers even certain weak forms of triviality "useless [for serious mathematics]" [11, p. 205] and then radicalized the claim to say that they are "catastrophic for mathematics" since it would render all calculation not only useless but also "meaningless" and "impossible" [13, p. 635]. He bases his case in a result by Dunn on real number theory according to which a single false equation makes the theory trivial without ever using *ex falso quodlibet*.

I want to argue, using some notions from category theory, that the step from triviality to meaninglessness, impossibility and uselessness is not as straight as Mortensen suggests. Paraconsistency theorists used to push the envelope and investigate the limits of logicality and mathematicality. This is an effort in the same, anti-incredulous stares spirit.[1] More than defending triviality in mathematics, which I do,[2] this work should be regarded as a plea for investigating it seriously, unless one is ready to employ an awkward double standard to avoid trivialism consisting in making exactly the same moves paraconsistentists railed against consistentists when it came to paraconsistency.

---

[1] And sometimes an incredulous stare means a lot of debate, because according to some, trivialism deserves no stare at all; cf. [5, p. 252].

[2] I am not alone on thinking that mathematics might have place for triviality. Priest in [15] considers models of arithmetic with (atomically) trivial objects in which, among other principles, neither the transitivity of '$=$' nor the substitutivity of identicals hold. That work was developed independently of Dunn's result, but the ideas serve to block it. Priest models are examples that no mathematical catastrophe needs to follow from a trivial arithmetical object. Nonetheless, I do not aim to compile here all the ways to block Dunn's result, so acknowledging Priest's work is enough for my purposes.

The plan of the paper is as follows. In the second section I make some terminological suggestions to give a better conceptualization of triviality in mathematics –for example, there are several reasons to prefer the label "atomic triviality" over Mortensen's "mathematical triviality"– and then I reconstruct Dunn's proof. In Sect. 3 I review Dunn's result on the light of degenerate toposes. A degenerate topos is an extremely simple mathematical universe (there is only one object up to isomorphism and its identity morphism) yet one can interpret logical notions in it, with the result that everything is true there due to the extreme simplicity of the universe. I will show that this implies that either one of the premises of Dunn's argument cannot obtain (from a point of view "external" to the universe) or that it obtains in calculations "internal" to such trivial universe and the theory associated, yet the calculations are possible and meaningful albeit extremely simple.

2 Dunn's Trivialization Result

2.1 *Terminological Preliminaries*

Mortensen abhors triviality in mathematics in a manner more thoughtful and cautious than most of his peers. But before assessing his claims, let me introduce some terminology. We can coin the term *C-trivial theory* for a theory T in which every sentence of a class C is true. Thus, the truth of all logically atomic sentences of a theory T is the *atomic triviality* of T. Mortensen calls "mathematical triviality" the atomic triviality of a mathematical theory, since for him mathematicality is closely tied to *functionality* –the validity of *Substitutivity of identicals* in atomic sentences, because that would be "what ensures that calculations can proceed" [13, p. 636]. Anything of what follows depends on the connections between mathematicality and functionality, so I stick to 'atomic triviality' because it is a more general case, neutral on mathematicality theses.[3]

Many cases of C-triviality coincide with triviality simpliciter under the presence of *ex falso quodlibet* (EFQ), but in general they do not and Mortensen in [11] presents some examples. However, he says that atomic triviality as has been just defined already is "catastrophic for mathematics" because, for example, a single false equation in real number theory is enough to produce atomic triviality without using EFQ. Moreover, Mortensen falls within the tradition of people (from Aristotle to Putnam through McTaggart and many others, see [17, Chap. 1]) who say that at least for some C, the C-triviality of T implies the meaninglessness of T: Mortensen

[3]At the eleventh hour prior to publication I was referred to two notions of relative triviality close to mine: In [18], negation-triviality (all the negations of a theory hold) is defined and called *quasi-triviality*, and in [7] 'quasi-triviality with respect to i and C' means that a contradiction with degree of complexity i implies all the formulas of a class C. Thanks to María del Rosario Martínez-Ordaz for the pointers.

says that calculations in such trivialized real number theory "would mean anything" and that they would not be "possible" or "useful" [13, p. 635].

In what follows, a version of the trivialization of real number theory through a single false equation without using EFQ is described.

2.2 Sketch of Dunn's Proof

Mortensen mounts his case on a result by Dunn which uses only some principles from high school algebra, notably principles about identity such as Transitivity and Substitutivity of Identicals, to show that real number theory becomes (logically) trivial by merely adding a single false equation. The argument runs as follows[4]:

– Rules of inference
Modus ponens (MP)
Universal instantiation (UI)
Uniform substitution (US)
– Principles for '='
Transitivity: For every x, y, z, if $x = y$ and $y = z$, $x = z$.
Substitutivity of identicals: For every x, y, if $x = y$ and $P(\ldots x \ldots)$, then $P(\ldots y \ldots)$.[5]
– Basic principles from real number theory
1. For every $x \in \mathbb{R}$, $(x - x) = 0$
2. For every $x, y, z \in \mathbb{R}$, if $x = y$ then $(x - z) = (y - z)$
3. For every $x, y, z \in \mathbb{R}$, if $x = y$ then $z(x) = z(y)$
4. For every $x \in \mathbb{R}$, $x(0) = 0$
5. For every $x, y, z \in \mathbb{R}$, there is a $w \in \mathbb{R}$ such that $z(x - y) = w$
– Rest of the argument
6. $a \neq b$, that is, that a and b are distinct real numbers; hypothesis.
7. $a = b$, a single false equation added to the theory to trivialize it.
8. $(a - a) = (b - a)$, from 2, 7, US and MP.
9. $0 = (b - a)$, from 1, 8 and *Substitutivity of identicals*.
10. $z(0) = z(b - a)$, from 3, 9, US and MP.
11. $0 = z(b - a)$, from 4, 10 and *Substitutivity of identicals*.
12. $0 = w$, from 5, 11 and *Substitutivity of identicals*.

Whether by *Transitivity* or the *Substitutivity of the identicals*, every real number equals any other. This is not even yet enough for triviality simpliciter, but only for atomic triviality.[6]

---

[4]I follow the version in [13, p. 635]; there are other, more intricate versions in [10, Chap. 6] and [11, Theorem 2].

[5]Actually, all what is needed is the "functional" version of this principle, that is, when $P(\ldots\tau\ldots)$ is a sentence free of logical connectives. The "transparent" version is when $P(\ldots\tau\ldots)$ is *any* sentence. Classically, the functional and the transparent versions of the principle are equivalent, but in general they are not in inconsistent mathematics. See [10, Chaps. 1 and 2] and [11].

[6]The extra resources to get full triviality are described in [11, p. 205].

Note that real number theory is then atomically trivialized by the addition of a single false equation, without using EFQ. I think this can be granted. However, my claim is that something else is needed to say that atomic triviality is "catastrophic for mathematics" besides the fact that it equals any real number to each other.

In the next section I will show a case of a fully, not only atomical, trivial mathematical universe where either premise 6 in Dunn's argument cannot obtain, or it can obtain but calculations are not "meaningless", "impossible" or "useless" and therefore no "catastrophe" follows from triviality.

3 Degenerate Toposes and Dunn's Result

3.1 Categorial Preliminaries

Categories are a kind of mathematical universe of objects and connections between them –their *morphisms*– satisfying very general conditions, like composability and associativity of morphisms as well as identity morphisms for every object – connections of objects with themselves.[7] An *isomorphism* is a morphism i between two objects X and Y with a morphism i^{-1} from Y to X such that the composition of i and i^{-1} is the identity morphism for X. Two objects are said to be *isomorphic* if there is a an isomorphism between them.

A central feature of standard category theory is that it is structural in the sense that each object in a category provably has all the same properties as any object isomorphic to it. For example, the defining property of a singleton is *having only one element*. Clearly, in usual set theories there are many singletons, but in a categorial set theory each singleton has only the properties that all of them have in common, so any of them can be denoted by the same sign, say '**1**', and speak as if there were only one of them.[8]

Toposes are categories with extra structure which allow for the interpretation of set-theoretical notions and hence of significant parts of mathematics, some of them even as much as ZFC. I do not need all the details of topos theory here, but only some aspects presented in a rather informal way that convey the main logical ideas. One of the crucial features of toposes is that there is a truth value ν which satisfies the

Comprehension axiom: The proposition $f(x)$ about an element x of a domain O is ν if and only if x belongs to the part M of the Os which are f's.

[7] Clear introductions to category theory in general, and topos theory in particular, can be found in [8].

[8] For those who might wonder of a definition in terms of objects and morphisms: a *terminal object* in a category $\mathbf{C}$, denoted '$\mathbf{1_C}$', is an object such that for any object X there is exactly one morphism from X to $\mathbf{1_C}$. The dual notion, *initial object*, denoted '$\mathbf{0_C}$', the categorial version of an empty set, is an object such that for any object X there is exactly one morphism from $\mathbf{0_C}$ to X.

The usual reading of this is that ν is the value $true$ and so M would be the *extension* of the predicate f. *Propositions* have the form "(An element) a belongs to the part M of (an object) O".[9]

This allows defining logical notions like $false$ and n-ary connectives of different orders in a way that a topos comes with an *internal logic*. This internal logic is internal in two very important senses. First, it is internal because it is defined using only the resources of the topos or mathematical universe in question; second, it is internal because it is the right logic to reason about the topos in question since it is determined by the definition of its objects and morphisms: Using a different logic to reason about them would alter their defining properties and thus it would not be a logic at all for the intended objects and their morphisms.

In short, truth values in such internal logic have the following features implied by the **Comprehension axiom** and the characteristics of any topos:
(IL1) Truth values form a partial order, i.e. for every values p, q and r:
(IL1a) $p \leq p$
(IL1b) If $p \leq q$ and $q \leq p$ then $p = q$
(IL1c) If $p \leq q$ and $q \leq r$ then $p \leq r$
(IL2) There is a truth value called $true$ with the following property:

$$\text{For every } p,\ p \leq \mathit{true}$$

(IL3) One can define a truth value called $false$ that has the following property:

$$\mathit{false} \leq \mathit{true}$$

and

$$\text{for every } p,\ \mathit{false} \leq p$$

(IL4) Rather than implied by the categorial data, the traditional, "Tarskian", notion of logical consequence is assumed:
Let '$p \models_{\mathcal{E}} q$' denote that q is a logical consequence of p in a topos $\mathcal{E}$, i.e. that whenever p is $true$ in $\mathcal{E}$, so is q. Equivalently: if q is not $true$, p neither is. $\models_{\mathcal{E}} p$ means that p is $true$ in $\mathcal{E}$.

Nothing in the above rules out a mathematical universe where the following hold:
(T1) For every p, $p = true$
(T2) $true = false$
(T3) For every p, $\models_{\mathcal{E}} p$
(T4) By (T1) and (T2), $p = true$ and $p = false$, for every p
These conditions are satisfied in a mathematical universe where all the objects are isomorphic, so for all practical purposes it can be said that there is only one object, D, and only one morphism, d, that must be with D itself. No element a and no part M of

[9]But there is an alternative, dual reading of ν as $false$, so M would be rather the *anti-extension* of the predicate f, etc. See [10, Chap. 11].

D can make the propositional function "x belongs to the part M of D" distinct from $true$, because D is the only object, it has no proper parts, and all of them is included in itself, so to speak. Thus, every propositional function (the only one expressible given the characteristics of this universe) is satisfied by every element of D –which is just D itself– and every proposition –for practical purposes, only one, since all propositions turn out to be equivalent given the characteristics of this universe– is $true$ –the only truth value given the characteristics of this universe–.

This goes further. As I have mentioned, a topos is a category which allows for the interpretation of set-theoretical notions. Thus, one has in it general categorial versions of, say, Cartesian binary products, disjoint unions or power sets. D and d are enough for a degenerate topos to satisfy the definitions of all these notions, so in a degenerate topos, a singleton is an empty set[10]; a Cartesian product is a disjoint union and a power set and so on.

3.2 Dunn's Result on the Light of Degenerate Toposes

A topos is a model of a set theory satisfying at least the axioms *Extensionality*, *Empty set*, *Pairs*, *Unions*, *Power-set*, *Foundation* and *Separation*. But a set theory restricted to these axioms is not enough for doing even some basic mathematics. For example, it does not assure that we have an object able to support recursive functions; for example, an object with infinite elements. One can introduce an axiom of *Infinity* (saying that there is a set with infinite members) in order to achieve that.

The categorial version of such an axiom is as follows. Let $\mathbf{C}$ a category with a terminal object $\mathbf{1}$. It is said that $\mathbf{C}$ has a natural numbers object (NNO for short) if in $\mathbf{C}$ there is an object N, a morphism $o:\mathbf{1}\longrightarrow N$ and a morphism $s:N\longrightarrow N$ with the following property: For every object X of $\mathbf{C}$, if there are morphisms $o':\mathbf{1}\longrightarrow X$ and $s':X\longrightarrow X$ then there is a unique morphism $h:N\longrightarrow X$ in $\mathbf{C}$ such that $(h\circ o)=o'$ and $(s'\circ h)=(h\circ s)$, i.e. such that makes the diagram below commutative:

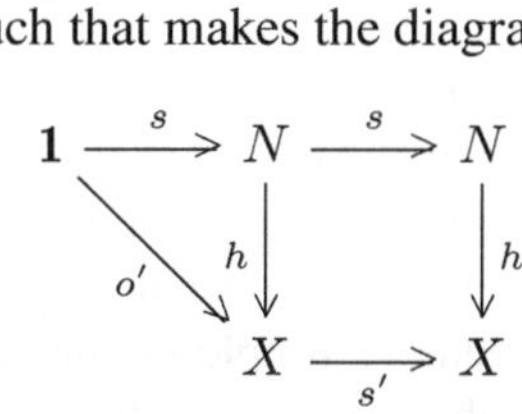

In any topos with a NNO, it is also possible to repeat the usual construction of the integers, the rationals, and then finally the real numbers; one thus obtains $\mathbb{R}$ in those categories. Actually, one can define a real numbers object (RNO) in any category with sufficient structure, not necessarily a topos. Then one can prove that an RNO exists in any topos with an NNO and in some other situations. I will not give the details of the categorial construction of the reals, this can be found in [6, Chap. 6];

[10] In fact, the usual non-degeneracy axiom states that terminal and initial objects are not isomorphic.

for my purposes it is enough to point out that a degenerate category supports a NNO and then also a RNO, so it makes sense to evaluate Dunn's proof in this setting.

Consider a degenerate topos viewed from the outside or "externally".[11] It consists of only one object and one morphism, so there cannot be false equations in it: There is no x in the topos such that it could be different from d. Thus, from an external point of view, Dunn's argument is unsound, for it requires in step 6 a distinction between objects that cannot be obtained.

Internally, the proof is sound, and the conclusion is just a complex way of saying that in a degenerate topos there is but a single number. Are calculations in such a degenerate mathematical universe *impossible*? Not really; all of them can be done, and very simply. Are calculations there *meaningless*? It depends. If one requires that some sentences mean something different from any other sentences, those calculations are meaningless. If one only requires that they mean something, they are not meaningless: Every sentence in the proof, in particular $a = b$ and $a \neq b$, means the same thing as any other sentence in a degenerate topos, namely "a [d] belongs to the part M [D] of D". Are those calculations *useful*? Probably not in a universe like ours where (presumably) not all objects are isomorphic, but certainly they are useful *qua* calculations for those degenerate toposes. Actually, any other kind of calculations would distort what happens in a degenerate topos; its internal logic, the right logic for doing calculations within that topos, is trivial after all.

What then about the catastrophic character of triviality in mathematics? Perhaps it is catastrophic *practically* for us, living beings in a universe which presumably is not a degenerate one. Even it might not be very interesting neither mathematically nor philosophically,[12] but there are reasons to think that there is nothing purely logically or purely mathematically catastrophic about triviality in mathematics.

4 Summary and Conclusions

Mortensen has put forward one of the few explicit, and one of the most compelling, cases against triviality in mathematics. He says that even some weak form of triviality is "catastrophic" because it makes calculations impossible, meaningless and useless. Apparently, a proof by Dunn on real number theory shows so. I have argued that degenerate toposes, certain extremely simple mathematical universes consisting of just one object and just one morphism (up to isomorphism) show that there is nothing catastrophic in Dunn's result for those universes. Thus, from a more general point of view, triviality makes perfect sense in its appropriate domain.

---

[11] The contrast between internal and external perspectives of a mathematical universe are already familiar in standard set theory. For example, up to some ordinal in the cumulative hierarchy one can see from the outside that there are infinite elements in the universe, but within the universe (up to that rank) those elements do not form yet a set, so within the universe there is no infinite set yet.

[12] Although I think it is, and this paper would be an argument for that, but I will not press this point further. See [3] for an example of how interesting things become for Platonists when triviality is taken a bit more seriously.

References

1. Beall, J. C. (1999). From full blooded platonism to really full blooded platonism. *Philosophia Mathematica*, *7*(3), 322–325.
2. Carnielli, W. A., & Coniglio, M. E. (2013). Paraconsistent set theory by predicating on consistency. *Journal of Logic and Computation*.
3. Estrada-González, L. Through full-blooded platonism, and what paraconsistentists could find there. *Logique et Analyse*, *59*(235), 283–300.
4. Kabay, P. (2010). *On the plenitude of truth. A defense of trivialism*. Germany: Lambert Academic Publishing.
5. Kroon, F. (2004). Realism and dialetheism. In G. Priest, J. C. Beall, & B. Armour-Garb (Eds.), *The law of non-contradiction. New philosophical essays* (pp. 245–263). Toronto: Oxford Clarendon Press.
6. Mac Lane, S., & Moerdijk, I. (1992). *Sheaves in geometry and logic: A first introduction to topos theory*. Heidelberg: Springer.
7. Martínez-Ordaz, M. del R. (2010). Gödel and paraconsistency: The road to non-triviality. Bachelor thesis, Universidad Veracruzana, Mexico (In Spanish).
8. McLarty, C. (1995). *Elementary categories. Elementary toposes*. Toronto: Oxford Clarendon Press.
9. Mortensen, C. (1989). Anything is possible. *Erkenntnis*, *30*(3), 319–337.
10. Mortensen, C. (1995). *Inconsistent mathematics*. Kluwer mathematics and its applications. Berlin: Kluwer Academic Publishers.
11. Mortensen, C. (2000). Prospects for inconsistency. In D. Batens, C. Mortensen, G. Priest, & J.-P. Van Bendegem (Eds.), *Frontiers of paraconsistent logic* (pp. 203–208). Research Studies Press.
12. Mortensen, C. (2005). It isn't so, but could it be? *Logique et Analyse*, *48*(189–192), 351–360.
13. Mortensen, C. (2007). Inconsistent mathematics: Some philosophical implications. In A. Irvine (Ed.), *Philosophy of mathematics* (pp. 631–649). Amsterdam: North Holland.
14. Priest, G. (1998). To be and not to be –that is the answer. On Aristotle on the law of non-contradiction. *Philosophiegeschichte und Logische Analyse*, *1*(1), 91–130.
15. Priest, G. (1998). The trivial object and the non-triviality of a semantically closed theory with descriptions. *Journal of Applied Non-Classical Logics*, *8*(1–2), 171–183.
16. Priest, G. (2000). Could everything be true? *Australasian Journal of Philosophy*, *78*(2), 189–195.
17. Priest, G. (2006). *Doubt truth to be a liar*. Oxford: Oxford University Press.
18. Robles, G. (2008). Weak consistency and strong paraconsistency. In J. María Díaz Nafría & F. Salto Alemany (Eds.), *Actas del I Encuentro Internacional de Expertos en Teorías de la Información. Un Enfoque Interdisciplinar* (pp. 117–131). León, España: Universidad de León.
19. Weber, Z. (2012). Review of 'inconsistent geometry', by Chris Mortensen. *Australasian Journal Philosophy*, *90*(3), 611–614.

On Gluts in Mathematics and Science

Andreas Kapsner

Abstract This essay investigates what role truth value gluts, statements that are both true and false, might play in accounting for inconsistencies in mathematics and science. More specifically, this essay asks whether truth value gluts should be *designated* values in such applications. Up to now, gluts have virtually always been designated, but I try to show that this might not be the best way to treat them.

Keywords Truth value gluts · Paraconsistency · Many valued logic · Designated value · Early calculus · Age of the earth · Darwin · Kelvin

1 Introduction

In this essay, I want to investigate what role truth value gluts, statements that are both true and false, might play in accounting for inconsistencies in mathematics and science. More specifically, this essay asks whether truth value gluts should be *designated* values in such applications. It is a special case of a general question that has been on my mind for a while, and which I have discussed for other special cases elsewhere[1] and plan to investigate further in the future. That question is simply: Should gluts be designated? There is a broad consensus that it can be uniformly answered in the positive, an answer I came to doubt.

Indeed, at some point in thinking about this, I thought that a case could be made for the opposite claim, that gluts should *never* be designated. But that now seems doubtful to me, as well, as some applications of gluts seem to push in the one, others to pull in the other direction. And some, like the ones I am about to discuss, seem much harder to decide than others. In case saying that makes you wonder how much rationale there is for asking the question in the first place, I will begin this

[1] See [12, 17]. [NB: *Pietz* was my name prior to marriage].

A. Kapsner (✉)
Munich Center of Mathematical Philosophy, Munich, Germany
e-mail: andreas.kapsner@googlemail.com; Andreas.Kapsner@lrz.uni-muenchen.de

H. Andreas and P. Verdée (eds.), *Logical Studies of Paraconsistent Reasoning in Science and Mathematics*, Trends in Logic 45,
DOI 10.1007/978-3-319-40220-8_6

investigation in the next section by giving an example for an application where I think the answer is clearly that gluts should *not* be designated.[2]

Before we look at that example, we should first determine what the question means in the first place. What does it mean to say that a value is designated?

First of all, the notion of a designated value is a technical device to define consequence in many valued logics. A valid inference in such a logic is one in which the conclusion must take a designated value whenever all of the premises do.

Philosophically, that is not yet very satisfying. As it happens, I've found surprisingly little discussion in the literature that addresses the question what designation *means*. I think that one of the best discussions of this questions is still the one in Dummett's famous paper *Truth*.[3] It is not, however, completely straightforward to apply his thoughts to the problem of truth value gluts, and I leave a detailed investigation for another day.

I will, nonetheless, work with an idea that is inspired by his discussion. According to this idea, the important question to answer when we ask whether a value is designated or not is this: What is the practical point of designation? What can you do with a statement that has a designated value that you can't do with a statement that has an undesignated one?

The answer to this I take from Dummett, as well: First and foremost, statements with designated values are the ones that we can correctly *assert*. Beyond that, these statements are such that we would be right in basing our actions and decisions on them. That much will do for the purposes of this essay. Our question, then, is this: When we meet a glut in science or mathematics, would we be right to assert it? Would we be right to base our decisions (e.g. about how to implement the scientific ideas in technology) on what a glutty statement tells us?

2 Epistemic Truth Values

To set up my discussion, let me quickly review something I have said elsewhere[4] about Belnap's interpretation of First Degree Entailment,[5] as it will not only help make my point, but also illustrate what undesignated gluts might come to logically. First Degree Entailment (FDE) has four truth values ($\mathcal{T}$rue, $\mathcal{F}$alse, $\mathcal{N}$either and $\mathcal{B}$oth) and can be characterized by the following well known lattice:

[2]For the opposing view that gluts should categorically be designated, see almost any author but me and my collaborators who writes about gluts; however, the topic is seldom even addressed explicitly. As I just stated, there are also applications in which *I* think gluts must be designated in order to do any significant work, and the clearest examples of those seem to me to be analyses of inconsistent art [15] and inconsistent fictions [16, 18].

[3]Reference [8].

[4]See [12, 17].

[5]Reference [3, 4].

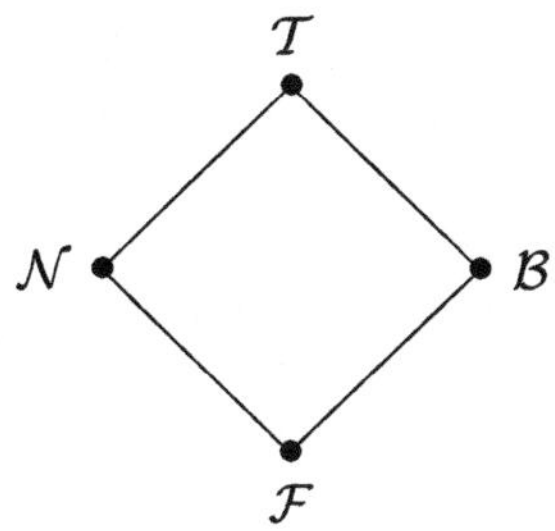

Conjunction is the *meet* (the greatest lower bound), disjunction the *join* (the least upper bound) and negation is an operator that flips $\mathcal{T}$ and $\mathcal{F}$ but is a fixed-point operator for $\mathcal{B}$ and $\mathcal{N}$.

Belnap thought of the four values as *epistemic* truth values, useful for dealing with information that a reasoning agent has received. They are contrasted with *ontological* truth values, roughly a substantive and realist kind of truth. I believe that we're much more likely to find use for epistemic truth values in accounting for inconsistencies in science, and the examples below will bear that out. For the time being, though, to get an idea what epistemic truth values and epistemic gluts in particular might be, think for example of the evidence at a trial. There can be conclusive-seeming evidence for a claim, against it, or no evidence either way. There is, however, also the fourth option of having convincing evidence *for as well as against* a claim. Think of two very reliable witnesses that one would not hesitate to believe, were it not for the fact that their testimonies contradict each other. The jurors will have to reason with what is given to them, and FDE is one suggestion for such reasoning.

Another one of the more concrete interpretations that Belnap offered is to think of the truth values as recording information that has been collected in a data base:

$\mathcal{N}$: The computer has received no information pertaining to the statement

$\mathcal{F}$: The computer has received the information that the statement is false

$\mathcal{T}$: The computer has received the information that the statement is true

$\mathcal{B}$: The computer has received the information that the statement is true and the information that it is false

The usual choice for designated values is $\mathcal{T}$ and $\mathcal{B}$, as they both contain some truth. But think about the examples for a bit, and think whether we should really assert or base our actions on statements with the value $\mathcal{B}$. If the evidence strongly suggests that "The defendant is guilty" should receive value $\mathcal{B}$, should the jurors assert it and thereby condemn the defendant?

Or suppose you are riffling through customer reviews in that big data base we stumble through every day, the internet. One of two products receives five star reviews throughout, while the other gets an even mix of five and zero stars. Supposing you have only these reviews and no more time for further inquiries, would it be rational for you to buy the second product because "This product is excellent (five stars)" receive value $\mathcal{B}$?

It seems clear to me that the answer to these questions is "No", and I've found in the past that the thumbnail sketch of an argument here was enough to convince many I put it to.[6]

The upshot is clearly that value $\mathcal{B}$ should not be designated. Designating only $\mathcal{T}$ gives us a perfectly good consequence relation, though for the details I refer you to my previous work. One thing that is noteworthy, though, is that not designating the glutty value results in the loss of paraconsistency, as it is impossible to construct a counter model for $A \wedge \neg A \models B$.[7]

3 The Early Calculus

Now that we've seen an example of undesignated gluts, let us turn to scientific and mathematical theories. I want to make a rough and ready distinction between two possible sources of inconsistencies in scientific or mathematical theorizing. The first is when we are dealing with a *single* theory that is found to be inconsistent, but which we wish to work with for one reason or other. I will discuss one of the classic examples of this in this section, the early calculus. The other kind of inconsistency I want to discuss arises from *two different* theories that have no or little overlap in content, but that, together with commonly shared auxiliary assumptions, lead to contradictions. My example for this will be the question of the age of the earth, as it was addressed by biologists and physicists in the second half of the 19th century. I will turn to this in the next section; first, the calculus.

When the early calculus was developed by Newton and Leibniz, it was operating on infinitesimally small quantities (dx) that had to be assumed to be different from 0 at some points in the calculation (because they were the denominators of fractions) and equal to 0 at other points (because they were dropped from sums, as in $2x + dx = 2x$).

Now, it has always been obvious that it won't do to add axioms of the form $dx = 0$ and $dx \neq 0$ to a suitable paraconsistent mathematical theory. This would still allow disastrous derivations as these:

$dx = 0$
$2dx = dx$
$2 = 1$

This suggests that simply treating $dx = 0$ as a glutty statement and treating the statements of the calculus with a logic in which gluts are designated will not be enough for a credible rational reconstruction. One of the best known attempts to provide such a reconstruction is the chunk-and-permeate strategy proposed by Brown and Priest

---

[6]For more detailed discussion, see [12, 17].

[7]As is customary, I understand a logic to be paraconsistent iff a contradiction does not logically entail an arbitrary statement. In the logic at hand there is, however, a counter model to $(A \wedge \neg A) \vee (B \wedge \neg B) \models C$, which should give a sense that we are dealing with a somewhat interesting notion of logical consequence here; it might be not too far off to see this feature as some deviant form of paraconsistency. Again, see [12, 17] for details.

[5]. In a nutshell, their idea is that reasoning with the calculus involved partitioning the assumptions into consistent chunks, an idea that had been around before them in the writings of non-adjunctive logicians. However, unlike them, Brown and Priest allow for some controlled information exchange between the chunks, taking care that no outright contradictions ensue.

Brown and Priest don't speak of gluts in their paper, and it is not clear to me how appropriate it would have been if they had. One could, I believe, maintain that their approach signals a move away from gluts as a useful device to account for the early calculus.

But let us go back and think about the following: Has the problem with the simple idea to add gluts to the picture maybe been that they were by default taken to be *designated* values? If they were taken to be undesignated values, disastrous deductions like the above would not go through. On the other hand, it would seem that many of the desirable deductions won't go through, either. It seems that you act on the belief that $dx = 0$ when you drop it from a sum, and act on the belief that $dx \neq 0$ when you divide by dx.

Peter Vickers in his book *Understanding Inconsistent Science*[8] argues that this is not the case, though. He makes a distinction between the *algorithmic level* and the *level of justification* for the calculus. He points out that for practical purposes, the "theory" of the early calculus was not a set of propositions, but rather a (sort of an) algorithm. The calculations that seemed to suggest that dx is equal to zero appear at very specific points in the procedure, while the ones that seem to suggest that dx is different from zero at others. As the calculations were reliably leading to the right results, it was quite possible to use the algorithm without a care about the existence of a possible *justification* for the procedure. Leibniz, for example,

> often stressed the pragmatic utility of his techniques, and how they could be exploited by mathematicians without their having to trouble themselves with foundational problems.[9]

One of those foundational problems would surely be to answer whether or not $dx = 0$. If one followed this advice, as many seem to have done,[10] then, Vickers claims, there was no need to attribute to them commitments to inconsistencies or gluts in any way at all.

But of course people were not *really* that uninterested in the justification of the calculus. They did not say *nothing* about it and accepted it as *just* a free floating algorithm. Maybe it is better to rationally reconstruct them as realizing that the calculus implied some statements to be gluts, but not asserting those glutty statements. To still be able to make use of the algorithm, maybe it is useful to recast the rules in a way that disowns the burden of proof: Instead of, e.g., "If you can assert that

[8]Reference [21].

[9]Reference [1], p. 20.

[10]A particularly nice quote expressing this sentiment that Vickers has found is this, by Oliver Heaviside: "Shall I refuse my dinner because I do not fully understand the process of digestion? No, not if I am satisfied with the result. (...) First, get on, in any way possible, and let the logic be left for later work" (Quoted from [21] p. 157).

$dx = 0$, drop dx from sums", adopt the rule "Drop dx from sums, unless you have reason to assert that $dx \neq 0$."

To be sure, all this does not amount to an ultimately satisfying state to be in. Berkeley would probably not have been silenced in his criticism by the retort that the glutty ghosts of departed quantities shall not be spoken of. An epistemic glut is always something that one hopes will go away with time, and in the case of the calculus it did. Note also that the employer of a chunk-and-permeate strategy isn't in an ultimately satisfying situation, either. As Priest wrote:

> If [a contradiction] is handled by a chunking strategy, then the theory is not a candidate for the truth. If α is true and $-\alpha$ is true, then so is their conjunction. If a theory refuses to allow this move then the theory cannot be correct, and we know this.[11]

4 Darwin and Kelvin on the Age of the Earth

I now propose to look in more detail at an example mentioned in Priest (2002) for inconsistencies between *different* theories. It concerns a very interesting episode in the history of science, the debate between biology and physics that took place in the second half of the 19th century. More precisely, it was a debate between evolutionary biology and thermodynamics, championed, respectively, by Darwin and Lord Kelvin[12] and their followers, and it went on for 60 years.

The point at which these theories were seen to be incompatible was the question of the age of the earth. (Note that this question is neither one of evolutionary biology nor of thermodynamics *per se*.) Here is the problem: On the basis of the new ideas in thermodynamics and some auxiliary assumptions and estimates, Kelvin calculated that the earth was probably around 100 million years old, maybe as young as 20 million years, but certainly not older than 500 million years. This was not enough time for Darwinian evolution.

Darwin had introduced the idea of evolutionary progress by random mutation and mechanical selection. This supplanted Lamarck's notion of progress by striving parental generations that passed on the fruits of their adaptive toils to their offspring, such as the giraffe that spends its life stretching for ever higher leaves and finally produces offspring with a slightly longer neck. Even Lamarck's rather bee-lining theory might not have worked in such a short time frame as was allowed by Kelvin; Darwin was certainly not able to entertain the plausibility of his theory of mechanistic progress under such a short estimate of the age of the earth.

He had been working on his theory of evolution relying on the assumptions of geologists of the early 19th century, such as Charles Lyell. Lyell held that the earth was very old, maybe indeterminably so. This was a slightly more cautions version of James Hutton's earlier view, who thought that the earth was cyclically reforming

[11]Reference [19], p. 126.

[12]At the beginning of the debate, Kelvin had not yet become a Lord and was known as William Thomson.

itself and thus eternal. The latter is an especially interesting extreme, given that relatively shortly before, the best guess at the age of the earth was a very precise calculation by Archbishop Ussher, based on biblical exegesis: The earth was created on October 22nd at 6pm in the year 4004 BC.[13] One can see the following scientific debate as a fascinating process of zeroing in on the present estimate of 4, 5 billion years that started at these extreme points.[14]

As is evident from the present estimate, Kelvin's calculation did not stand up. He had not, indeed could not have, taken a new source of heat energy into account. This was nuclear radiation, which was only discovered towards the end of his life. It is maybe with some hindsight bias that modern authors claim that he "entered the debate with all the arrogance of a newly established 'science of the century', namely the recently drafted laws of thermodynamics."[15] Though it is true that he did not mince his words, would Kelvin have appeared all that arrogant to us if his estimate of the age of the earth would have panned out?

Before Kelvin entered the debate (whether in an arrogant or deservedly self-confident manner), Darwin was himself maybe a bit overconfident in voicing his support of the old age of the earth. He wrote in *On the Origin of Species*:

> He who can read Sir Charles Lyell's grand work on the Principles of Geology, which the future historian will recognize as having produced a revolution in natural science, yet does not admit how incomprehensibly vast have been the past periods of time, may at once close this volume.[16]

To illustrate the immense time scales of geological processes, he himself contributed a back-of-the-napkin calculation of the length of a particular episode of erosion in a valley in the south of England. The result of his quick calculation was that the process "must have required 306,662,400 years; or say three hundred million years".[17]

However, he very soon come to regret taking such a bold stance, and he came to see Kelvin's attack as one of the gravest problems for his theory.[18] Even backed, as he was, by formidable minds such as T.H. Huxley's, he could not defuse that attack.

The long debate is a most fascinating one,[19] with many surprising turns. For example, Kelvin, notwithstanding his complete lack of sympathy for evolutionary ideas, at one point offered a way out of his own vice-like grip: Life might not have been developing on earth all along, but might have been brought in from elsewhere on a meteorite. Instead of grasping that unexpected lifeline, Huxley made fun of it:

---

[13]Before you smirk, remind yourself that these kinds of estimates are still around today.

[14]An interesting account of the beginning of this debate up to Lyell is Gould (1987), who sets out to debunk what he thinks is a myth, the claim that Hutton and Lyell were making any kind of *scientific* progress over the theological arguments. In his view, early geology was in this question as unencumbered by actual empirical evidence as the Mosaic speculations had been.

[15]Reference [20] p, 213.

[16]Quoted from [7], p. 301.

[17]Quoted from [7], p. 303.

[18]See [7], p. 303.

[19]Indeed, it has even been turned into a stage play [20].

> "God almighty sitting like an idle boy at the sea side and shying aerolites (with germs), mostly missing, but sometimes hitting a planet!"[20]

Rather than latching on to this idea, Huxley tried to make do with the short time frame handed to him by physics, an attempt that did not convince many critics.[21]

To sum up this long debate, Priest is unsurprisingly right that no one made assertions that would be well modeled by designated gluts:

> [G]iven the dispute about the age of the earth at the end of the 19th century, no one conjoined the views that the earth was hundreds of millions of years old, and that it was not, to infer that the earth really had a contradictory age.[22]

Might we better interpret them by employing undesignated gluts? Darwin himself comes to mind. He *did* strike out passages as the following from later editions of *On the Origin of Species*:

> [I]t is highly important for us to gain some notion, however imperfect, of the lapse of time. During each year the land and water have been peopled by hosts of living forms. What an infinite number of generations, which the mind cannot grasp, must have succeeded each other in the long roll of years.[23]

However, he in turn added statements like these:

> With respect to the lapse of time not having been sufficient since our planet was consolidated for the assumed amount of organic change, and this objection, as urged by Sir William Thompson, is probably one of the gravest as yet advanced, I can only say, firstly that we do not know at what rate species change as measured in years, and secondly, that many philosophers are not yet willing to admit that we know enough of the constitution of the universe and of the interior of our globe to speculate with safety on its past duration.[24]

This looks as if Darwin wanted to suggest the statement "The earth is older than 500 million years" to be an epistemic *gap* rather than a glut. Dialectically, this might have been a more promising strategy, and in fact one that can be seen as having been vindicated from our point of view.

In fact, no one I could identify appears to have been in an epistemic situation that should be modeled by a glut. Every commentator, partisan or not, professed doubts about some of the assumptions; sometimes these doubts were balanced out to speak against both sides.[25] More often, they were concentrated on one side or the other of the debate, so as to warrant a definite stance on the question.

It would seem, then, that we would do justice to no one who publicly commented on the debate if we ascribed a glutty commitment to them. That certainly doesn't mean that such commitments couldn't be plausibly held at the time; indeed, it doesn't even mean that they *weren't* held by someone, it only means that such persons did not

---

[20]Cited from [11] p. 114.

[21]Reference [11], p. 115.

[22]Reference [19] p. 123.

[23]Quoted from [6], p. 311.

[24]Quoted from [6], p. 319.

[25]Such as in an article published in the Edinburgh Review, summarized in [7], p. 94.

speak up. As it happens, that would be just what the undesignated glut view would suggest to them, at least on the crucial question whether the earth was older than 500 million years.

You would have been pushed towards accepting this statement to be glutty if you were, at the time, a firm believer both in natural selection and thermodynamics,[26] and were additionally signing up to all the auxiliary assumptions (such as that the physicists knew about all of the possible sources of energy or that terrestrial life evolved in its entirety on planet earth and was not given a head start by, say, imported bacteria imported on a meteorite).

Compare this situation with the situation of the juror as described above, and the similarity should be conspicuous: The roles of the two credible witnesses are taken up by the two theories.

Now, what should we say about the question of designation? What kinds of firm assertions should you make about the age of the earth? Should you firmly state that, given your scientific convictions, the earth is and isn't older than 500 million years? Or should you rather refrain from making such assertions about the age of the earth?[27]

It seems clear, again just as in the case of the juror, that the latter option is the right one.

5 Conclusion

I hope to have shown that simply adding designated gluts to a logical semantics will not lead to any satisfying result in the two examples I discussed. Leaving the gluts undesignated, however, has some utility in accounting for what can rightly be asserted by someone who wants to reason with an inconsistent theory or a pair of incompatible ones. As far as this goes, it lends support to my general thesis that it should not be taken as a foregone conclusion whether or not gluts should be designated.

Undesignated gluts will take some time to get used to. The consequence relations they give rise to might look a little unusual, as is the case for the logic I described in Sect. 2. Also, they seem to push us toward some kind of non-monotonicity, at least pragmatically (the logics themselves might well be monotonic). We may find that things we thought were implied by our theories are actually gluts, and are thus

[26]Without, however, having skin in the game.

[27]One referee remarked that my discussion suggests a Popperian view in which these statements should count as unfalsified, and that such statements might be better modeled as gaps than as gluts. The present paper, as it were, operates on the assumption that gluts make sense in these settings, but in fact I am quite sympathetic to the comment (see [12, 14] is a paraconsistent account of Popperian science in which, to me at least, it seems neither fitting to speak of gluts nor of gaps). Relatedly, both referees remark that the possibility of undesignated gluts seems to significantly blur the line between gaps and gluts. This is especially true if one also allows for the possibility of *designated gaps* (and I think there is a place for those, too; see, again, [12]). I think the line is still discernible, but it might well be that one has to squint especially hard in the cases I discuss here.

actually *not* consequences of our theories. When we combine two theories, we may thus lose some statements that we had thought we could safely assert.

With a view to the name of this book, let me end on the following note: I mentioned in passing that this strategy will yield a logic that is *not* paraconsistent.[28] There is no danger in this, as you will not be led to assert every statement whatsoever because it logically follows from what you asserted earlier. After all, the gluts, though present, are not asserted.[29]

References

1. Arthur, R. (2008). Leery bedfellows: Newton and Leibniz on the status of infinitesimals. In U. Goldenbaum & D. Jesseth (Eds.), *Infinitesimal differences: Controversies between Leibniz and his contemporaries* (pp. 7–30). Germany: De Gruyter.
2. Badash, L. (1989). *The age-of-the-earth debate*. New York: Scientific American.
3. Belnap, N. (1977a). A useful four-valued logic. In J. Dunn & G. Epstein (Eds.), *Modern uses of multiple-valued logic* (pp. 8–37). Dordrecht: Reidel Publishing Company.
4. Belnap, N. (1977b). How a computer should think. In G. Ryle (Ed.), *Contemporary aspects of philosophy* (pp. 30–57). London: Oriel Press.
5. Brown, B., & Priest, G. (2004). Chunk and permeate, a paraconsistent inference strategy. Part I: The infinitesimal calculus. *Journal of Philosophical Logic*, *33*(4), 379–388.
6. Burchfield, J. (1974). Darwin and the dilemma of geological time. *Isis*, *65*, 301–321.
7. Burchfield, J. (1975). *Lord Kelvin and the age of the earth*. New York: Science History Publications.
8. Dummett, M. (1978). Truth. In *Truth and other enigmas* (pp. 1–25). Massachusetts: Harvard University Press.
9. Gould, S. J. (1987). *Time's arrow, time's cycle: Myth and metaphor in the discovery of geological time*. Massachusetts: Harvard University Press.
10. Hattiangadi, J. N. (1971). Alternatives and incommensurables: The case of Darwin and Kelvin. *Philosophy of Science*, *38*, 502–507.
11. Hellman, H. (1998). *Great feuds in science: Ten of the liveliest disputes ever*. New York: Wiley.
12. Kapsner, A. (2014). *Logics and falsifications* (Vol. 40). Trends in logic. Heidelberg: Springer.
13. Marcos, J. (2011). The value of the two values. In J.-Y. Béziau & M. E. Coniglio (Eds.), *Logic without Frontiers: Festschrift for Walter Alexandre Carnielli on the occasion of his 60th birthday* (Vol. 17, pp. 277–294). Tribute. London: College Publications.
14. Miller, D. (2006). Making falsification bite. *Out of error: Further essays on critical rationalism*. Burlington: Ashgate Publishing.
15. Mortensen, C. (1999). Peeking at the impossible. *Notre Dame Journal of Formal Logic*, *38*(4), 527–534.
16. Pietz, A. (2010). Tales of Explosion. In M. Pelis (Ed.), *The Logica Yearbook 2009*. Academy of Sciences of the Czech Republic.
17. Pietz, A., & Rivieccio, U. (2013). Nothing but the truth. *Journal of Philosophical Logic*, *42*(1), 125–135.

[28]Provided no other mechanism is invoked to secure paraconsistency. If for example, some gluts are undesignated, but some others remain designated, we can again construct counterexamples to Explosion.

[29]This essay has benefited greatly from the comments of two anonymous referees, as well as from discussion at the conference *Paraconsistent Reasoning in Mathematics and Science*. The research was financed by the research project "New Logics for Verificationism" PI 1082/1-102288852134 funded by the German Research Foundation.

18. Priest, G. (1999). Sylvan's box: A short story and ten morals. *Notre Dame Journal of Formal Logic*, *38*(4), 573–582.
19. Priest, G. (2002). Inconsistency and the empirical sciences. In J. Meheus (Ed.), *Inconsistency in science* (Vol. 2, pp. 119–128). Heidelberg: Springer.
20. Stinner, A., & Teichmann, J. (2003). Lord Kelvin and the age-of-the-earth debate: A dramatization. *Science and Education*, *12*, 213–228.
21. Vickers, P. (2013). *Understanding inconsistent science*. Oxford: Oxford University Press.

Contradictoriness, Paraconsistent Negation and Non-intended Models of Classical Logic

Carlos A. Oller

Abstract Given that, by definition, two statements are contradictories if and only if it is logically impossible for both to be true and logically impossible for both to be false, some authors have argued that the negation operators of certain paraconsistent logics are not "real" negations because they allow for a statement and its negation to be true together. In this paper we argue that the same kind of argument can be levelled against the negation operator of classical propositional logic. To this end, Carnap's result that there are models of classical propositional logic with non-standard or non-normal interpretations of the connectives, and that one kind of those interpretations violate the semantical principle of non-contradiction which requires of a sentence and its negation that at least one of them be false can be used. We ponder the consequences of these arguments for the claims that paraconsistent negations are not genuine negations and that the negation of classical logic is a contradictory-forming operator and we consider the arguments that challenge the conflation between negation and contradiction.

Keywords Classical negation · Paraconsistent negations · Contradictory-forming operators · Carnap's non-standard models of classical logic

1 Introduction

It is usually accepted in the literature that negation is a contradictory-forming operator and that two statements are contradictories if and only if it is logically impossible for both to be true and logically impossible for both to be false. These two premises have

C.A. Oller (✉)
Facultad de Filosofía y Letras, Departamento de Filosofía,
Universidad de Buenos Aires, Buenos Aires, Argentina
e-mail: carlos.a.oller@gmail.com

C.A. Oller
Facultad de Humanidades y Ciencias de la Educación, Departamento de Filosofía - IdHICS,
Universidad Nacional de La Plata, La Plata, Argentina

H. Andreas and P. Verdée (eds.), *Logical Studies of Paraconsistent Reasoning in Science and Mathematics*, Trends in Logic 45,
DOI 10.1007/978-3-319-40220-8_7

been used by Hartley Slater [13] to argue that paraconsistent negation is not a "real" negation because a sentence and its paraconsistent negation can be true together.

In this paper we claim that a counterpart of Slater's argument can be directed against the negation operator of classical logic. Carnap's discovery that there are models of classical propositional logic with non-standard or non-normal interpretations of the connectives will be used to build such an argument. One such non-normal valuation which can be added to the set of classically admissible valuations without altering the set of theorems or the set of valid consequences assigns *true* to every well-formed formula and, therefore, assigns a designated value to every formula and its negation.

We ponder the consequences of these arguments for the claims that paraconsistent negations are not genuine negations and that the negation of classical logic is a contradictory-forming operator. To this end, we follow the arguments that Dutilh Novaes develops in [4] to challenge the conflation between negation and contradiction.

2 "Genuine" and Paraconsistent Negations

Some authors have argued that the negation operators of certain paraconsistent logics—i.e. logics which do not validate the *ex contradictione quodlibet* rule (*ECQ*): $\{A, \neg A\} \models B$, for every A and B—are not "real" negations. Given that, according to them, a "genuine" negation is a contradictory-forming operator and two statements are contradictories if and only if it is logically impossible for both to be true and logically impossible for both to be false, the negations of those logics are not "real" negations because they allow for a statement and its negation to be true together.

In a much quoted paper Hartley Slater maintains that the negation of Graham Priest's [9] paraconsistent logic *LP* (*Logic of paradox*) is not a genuine negation because in the three-valued semantics for *LP* there are two designated truth values that count as being true: t (true only), and b (both true and false), and both A and $\neg A$ can receive the designated value b in *LP*. Ironically, some years earlier Richard Routley and Graham Priest [10] had directed a similar criticism against the negation operator of da Costa's paraconsistent logic C_1 and had concluded that da Costa's negation was merely a subcontrary-forming operator—i.e. that a sentence and its da Costa's negation cannot both be false though they may both be true—.

Slater maintains that the same line of reasoning can be applied to every paraconsistent system and concludes that, properly speaking, there is no paraconsistent negation. The following argument—Slater's argument against paraconsistent negations as reconstructed by Francesco Paoli [8]—summarizes the above:

(1) Contradictories cannot be true together.
(2) A sentence and its negation are contradictories.
(3) If L is a paraconsistent logic, then, in the semantics for L, there are valuations which assign both A and $\neg A$ a designated value, for some formula A.

(4) If A and B both receive a designated value, under some valuation v, in the semantics for L, then A and B can be true together according to L.
(5) In paraconsistent logics, A and $\neg A$ may not be contradictories (from (1), (3), (4)).
(6) Thus, paraconsistent "negations" are not negations (from (2), (5)).

It can be argued that Slater obtains an easy victory because he assumes that "real" negations are, by definition, contradictory-forming operators [1]. Instead of questioning this assumption, in what follows we present an argument that uses Slater's premises to conclude that, if we accept them, not even classical negation can be considered a "genuine" negation.

3 Classical Negation and Non-standard Models of Classical Logic

In this section we will argue that the same kind of argument that Slater directs against paraconsistent negations can be levelled against the negation operator of classical propositional logic. To this end, Carnap's result that there are models of classical propositional logic with non-standard or non-normal interpretations of the connectives can be used.

In his *Formalization of Logic* Carnap tried to solve what he called *the problem of a full formalization of (first-order) logic*, i.e. "whether—and, in what way—it is possible to construct a calculus (…) such that the principal logical signs can be interpreted only in the normal way" [2, p. 3]. After proving that the customary formalizations of first-order logic do not achieve full formalization he introduced a multiple-conclusion presentation of elementary logic that he claimed to fulfill that goal, even though in his review of Carnap's solution Alonzo Church manifested his scepticism and conjectured that "non-normal interpretations of the propositional calculus can be excluded only by semantical (as opposed to purely syntactical) rules" [3, p. 496].

Carnap proves that there exist sound bivalent valuations—with respect, for example, to the standard natural deduction rules for classical propositional logic—that do not conform to the classical truth tables for the connectives. One kind of non-normal valuations violate the semantical principle of non-contradiction, which requires of a sentence and its negation that at least one of them be false. Carnap proved that the only non-normal (sound) bivalent valuation of this type is the valuation $v_{\top}$ which assigns the truth-value t (*true*) to every formula, i.e. for every sentence A, $v_{\top}(A) = t$. Let V be the set of standard classically admissible valuations and V' an extended set of admissible bivalent valuations such that $V' = V \cup \{v_{\top}\}$. It is easy to show that these two different sets of admissible valuations determine the same consequence relation—in symbols, $\Gamma \models_{V} A$ iff $\Gamma \models_{V'} A$, for every set of formulas Γ and every formula A—and, therefore, the same set of logical truths and valid inferences.

The other kind of non-normal valuations violate the semantical rules that the negation of a false sentence must be true and that a disjunction is false if its disjuncts are both false. An example of a valuation of this second kind is the one that assigns the truth value *true* to those formulas which are theorems of classical propositional logic and *false* to those formulas which are not theorems of classical propositional logic.

The interest in Carnap's discovery of non-standard models for classical logic has recently been revived in relation with the inferentialist thesis that the meanings of the logical constants are completely determined by their introduction and elimination rules in a natural deduction system [7, 12]. But, as we will try to show in what follows, those non-standard models are also relevant for the discussion of the philosophically adequate characterization of metalogical notions—such as that of contradictoriness—and their relation with different kinds of negation.

Taking into account Carnap's results, it is possible to build the counterpart for Slater's argument against paraconsistent negations in the case of classical negation:

(1) Contradictories cannot be true together.
(2) A sentence and its negation are contradictories.
(3) There exists a (non-standard) sound and complete bivalent semantics for classical logic such that there are valuations in this semantics which assign both A and $\neg A$ the designated value, for every formula A.
(4) If A and B both receive the designated value, under some valuation v, in an adequate bivalent semantics for classical logic, then A and B can be true together.
(5) In classical logic, A and $\neg A$ may not be contradictories (from (1), (3), (4)).
(6) Thus, classical "negation" is not a negation (from (2), (5)).

4 Is Classical Negation a Contradictory-Forming Operator?

In order to ponder the consequences of Carnap's result for the case against classical negation as a contradictory-forming operator we need to fix the definitions of "negation", "contradictories"—the term "contradictories" allow us put into brackets the question about the kind of entities involved in the notion of contradiction—and "classical logic". The term "contradictories" used here allow us to postpone the question about the kind of entities that can be used to characterize the notion of contradiction. It has been pointed out that at least four different approaches to the notion of contradictories can be found in the literature [6]: semantic definitions in terms of possibility, truth and falsity; syntactic definitions in terms of form; pragmatic definitions in terms of assertion and denial; and ontological definitions in terms of states of affairs.

In his argument against paraconsistent negations Slater uses a semantic notion of contradictories and assumes that genuine negations are contradictory-forming operators. But it should be noted that the usual semantic definition of "contradictories"—

two statements (sentences, propositions, formulas) are contradictories if and only if it is logically impossible for both to be true and logically impossible for both to be false—does not involve the notions of negation or classical logic, two notions whose characterization is certainly problematic. As pointed out by Dutilh Novaes, the idea of negation as a contradictory-forming operator is a quite recent development in the history of logic and an examination of the history of this discipline shows that the syntactical notion of negation and the semantic notion of contradiction can be conceptually independent of each other. In fact, Novaes points out that the notion of contradiction in Aristotelian logic does not have a straightforward syntactical propositional counterpart because Aristotle's negation is a term-negation and, therefore, a non-propositional one. It is only in the twentieth century that the notion of negation as a contradictory-forming propositional operator has become the predominant one and its source can be found in Frege's notion of negation as a function that maps the True to the False and the False to the True. This concept of propositional negation as the syntactic counterpart of the semantic notion of contradictory propositions is clearly stated in Whitehead and Russell's *Principia Mathematica*:

> The Contradictory Function with argument p, where p is any proposition, is the proposition which is the contradictory of p, that is, the proposition asserting that p is not true. This is denoted by $\sim p$. Thus $\sim p$ is the contradictory function with p as argument and means the negation of the proposition p. It will also be referred to as the proposition not-p. Thus $\sim p$ means not-p, which means the negation of p. [15, p. 6]

Dutilh Novaes concludes that, given that most of the notions of negation that can be found throughout the history of logic are not contradictory-forming operators, Slater's argument is not sound because one of its premises is simply not true and, therefore, Priest's paraconsistent negation is, at least in principle, as genuine a negation as any other.

Dutilh Novaes defense of paraconsistent negations can be used, *mutatis mutandis*, to accommodate Carnap's non-intended interpretations of propositional logic that allow for a formula and its negation to be both true. Her point of view permits us to accept the following statement made by Slater: "...[Priest] tries to show that Boolean negation likewise involves an operator for which the truth of $\neg\alpha$ does not rule out that of α. But, even if this was true, it would merely show that Boolean negation was not a contradiction-forming operator ..." [14, p. 458]. Given the premises he accepts and taking into account the existence of Carnap's non-normal valuations, this would seem a sensible conclusion for Slater to draw with respect to classical logic. Nevertheless, if contradictory-forming negations are just one kind of (real) negations, the fact that classical negation is not a contradictory-forming operator does not oblige us to accept that it is not a genuine negation. And this because it is possible to assign both A and $\neg A$, for every formula A, the designated value t within a sound and complete bivalent semantics for a natural deduction presentation of classical logic.

Of course, one can try to circumvent Carnap's results by characterizing classical propositional logic as the logic determined by standard classical models—i.e. as the set of logical truths and valid inferences determined by those models—and classical negation as the contradictory-forming connective characterized by its standard bivalent truth-table. But this strategy seems to be a question-begging one: it assumes what needs to be proved, i.e. that Carnap's non-standard semantics is not a *bona fide* (bivalent) one for classical propositional logic. But, given that Carnap's non-standard models seem to provide such a semantics—because these models determine the same set of logical truths and valid inferences as standard classical models—the burden of proof lies with those who maintain that these results do not concern classical negation or classical logic. They must make explicit the difference—and the relevance of such a difference—between the logics determined by the standard and non-standard models that justify their stance, because otherwise their strategy would seem an ad hoc application of the advice "When you meet a contradiction, make a distinction."

It might be argued that even though Carnap's valuation $v_\top$ is unobjectionable from the point of view of a formal or pure semantics, it is not possible to provide a sensible informal or intuitive account of $v_\top$. If valuations are considered as descriptions of possible worlds or states of affairs, then $v_\top$ seems to commits us to a (weak) form of trivialism according to which there is a world where every sentence holds [5, 11]. However, it is debatable whether such a world can be discarded on purely logical grounds. But, be that as it may, even if we consider only those states of affairs in which at least one proposition is false, Carnap's second kind of non-normal valuations show that the natural deduction rules for classical propositional logic do not constrain us to accept that the classical negation operator is the syntactic counterpart of the truth function which maps truth to falsehood and falsehood to truth.

5 Conclusion

Carnap's non-standard models for classical logic have been mainly discussed in relation with the inferentialist conception of the meaning of the logical constants. But, as we have tried to show in this paper, those non-standard models are also relevant for the discussion of the relation of semantic notions such as contradictoriness and its relation with different (syntactic) notions of negation. In particular, we show that Slater's argument against paraconsistent negation, which assumes that a "genuine" negation is the syntactic counterpart of the notion of contradiction, can be directed, *mutatis mutandis*, against classical negation: in view of Carnap's results, if Slater's argument were a good one then neither paraconsistent negation nor classical negation would be "real" negations. But, as the conflation between propositional negation and contradiction is not a conceptual necessity, the genuineness of classical—and paraconsistent—negation can be defended but its contradictory-forming nature—at least, according to the usual semantic characterization of the notion of contradictoriness—is doubtful.

References

1. Béziau, J.-Y. (2006). Paraconsistent logic! (A reply to Slater). *Sorites*, *17*, 17–25.
2. Carnap, R. (1943). *Formalization of logic*. Cambridge: Harvard University Press.
3. Church, A. (1944). Review of Carnap's formalization of logic. *The Philosophical Review*, *53*, 493–498.
4. Dutilh Novaes, C. (2007). Contradiction: the real challenge for paraconsistent logic. In J. Y. Béziau, W. A. Carnielli, & D. Gabbay (Eds.), *Handbook of paraconsistency* (pp. 465–480). London: College Publications.
5. Estrada-González, L. (2012). Models of possibilism and trivialism. *Logic and Logical Philosophy*, *21*, 175–205.
6. Grim, P. (2004). What is a Contradiction? In G. Priest, J. C. Beall, & B. P. Armour-Garb (Eds.), *The law of non-contradiction: New philosophical essays* (pp. 49–72). Oxford: Oxford University Press.
7. Murzi, J., & Hjortland, O. T. (2009). Inferentialism and the categoricity problem: Reply to Raatikainen. *Analysis*, *69*, 480–488.
8. Paoli, F. (2003). Quine and Slater on paraconsistency and deviance. *Journal of Philosophical Logic*, *32*, 531–548.
9. Priest, G. (1979). The logic of paradox. *Journal of Philosophical Logic*, *8*, 219–241.
10. Priest, G., & Routley, R. (1989). Systems of paraconsistent logic. In G. Priest, R. Routley, & J. Norman (Eds.), *Paraconsistent logic: Essays on the inconsistent* (pp. 151–186). Philosophia: Munich.
11. Priest, G. (2006). *Doubt truth to be a liar*. Oxford: Oxford University Press.
12. Raatikainen, P. (2008). On rules of inference and the meanings of logical constants. *Analysis*, *68*, 282–287.
13. Slater, B. H. (1995). Paraconsistent logics? *Journal of Philosophical Logic*, *24*, 451–454.
14. Slater, B. H. (2007). Dialetheias Are Mental Confusions. In J.-Y. Béziau, W. Carnielli, & D. Gabbay (Eds.), *Handbook of paraconsistency* (pp. 457–466). London: King's College.
15. Whitehead, A. N., & Russell, B. (1910). *Principia Mathematica* (Vol. 1). Cambridge: Cambridge University Press.

From Paraconsistent Logic to Dialetheic Logic

Hitoshi Omori

Abstract The only condition for a logic to be paraconsistent is to invalidate the so-called explosion. However, the understanding of the only connective involved in the explosion, namely negation, is not shared among paraconsistentists. By returning to the modern origin of paraconsistent logic, this paper proposes an account of negation, and explores some of its implications. These will be followed by a consideration on underlying logics for dialetheic theories, especially those following the suggestion of Laura Goodship. More specifically, I will introduce a special kind of paraconsistent logic, called *dialetheic* logic, and present a new system of paraconsistent logic, which is dialetheic, by expanding the Logic of Paradox of Graham Priest. The new logic is obtained by combining connectives from different traditions of paraconsistency, and has some distinctive features such as its propositional fragment being Post complete. The logic is presented in a Hilbert-style calculus, and the soundness and completeness results are established.

Date: March 16, 2016.
The author was a Postdoctoral Fellow for Research Abroad of the Japan Society for the Promotion of Science (JSPS) at the time of submission, and now a Postdoctoral Research Fellow of JSPS. I would like to thank Holger Andreas and Peter Verdée for their encouragement and patience. I would also like to thank Diderik Batens, Filippo Casati, Petr Cintula, Michael De, Graham Priest, Dilip Raghavan, Greg Restall, Daniel Skurt, Heinrich Wansing and Zach Weber for their valuable suggestions, comments and discussions. Earlier versions of the paper were presented to conference *Paraconsistent Reasoning in Science and Mathematics* in Munich, *Workshop on Non-Classical [Meta]Mathematics* in Otago and seminars in Ghent, Munich, Singapore and Melbourne, and many thanks go to organizers and audiences. Finally, I would like to thank the referees for their helpful comments and suggestions which substantially improved the paper.

H. Omori
The Graduate Center, City University of New York, New York, NY, USA

H. Omori (✉)
Department of Philosophy, Kyoto University, Kyoto, Japan
e-mail: hitoshiomori@gmail.com
URL: https://sites.google.com/site/hitoshiomori/home

H. Andreas and P. Verdée (eds.), *Logical Studies of Paraconsistent Reasoning in Science and Mathematics*, Trends in Logic 45,
DOI 10.1007/978-3-319-40220-8_8

1 Introduction

Dialetheism is the metaphysical view, not restricted to any of the specific topics, that some contradictions are true. At first sight, dialetheism looks not tenable at all, due to the popular view that no contradictions are true, which is based on the Law of Non-Contradiction since Aristotle. However, some philosophers, such as Jc Beall, Graham Priest and Richard Routley (later Sylvan), have challenged the Law of Non-Contradiction, and defended dialetheism.[1]

In this paper, I wish to pave the path towards satisfactory examination of dialetheism in the context of foundations of mathematics which is one of the topics discussed by Priest in his celebrated '*In Contradiction*'. The motivation behind this project is rather simple: we want to see how much mathematics can be developed by keeping the following claim of Priest:

> I wish to claim that (Abs) and (Ext) are true, and in fact that they analytically characterise the notion of set. ([32, p.30])[2]

Now, in order to develop some formal theory for foundations of mathematics, we need to deploy an underlying logic which justifies the inferences made in the formal theory. In mathematics, the standard underlying logic is classical logic. However, if we buy the axioms for naive set theory, then we immediately face a contradiction. And in order to handle contradictions, we need an alternative system of logic. A family of non-classical logics that is capable of handling contradictions is known as paraconsistent logic. Priest has been known for claiming that systems of paraconsistent logic that contain classical negation are not appropriate for the purpose of developing dialetheic theories. However, this turned out to be not the case in general, as observed in [25], if one follows the suggestion of Laura Goodship in [15]. In brief, Goodship's suggestion is to take the material conditional and biconditional defined in terms of paraconsistent negation in formulating dialetheic theories. If one can have classical negation as well, then this means that there are even more candidates than before in deciding which logic should be the underlying logic, which at least needs to be paraconsistent for the present purpose.

But what are the desiderata for paraconsistent logic? As is well known, paraconsistent logics are independent of dialetheism in general. More specifically, dialetheism calls for paraconsistent logic, but not the other way around. Indeed, at the modern origin of paraconsistency, dialetheism was *not* the reason to develop paraconsistent logics. And based on various reasons, infinitely many systems of paraconsistent logic have been developed. So far, the only condition for a logic to be called paraconsistent is the following:

$$A, {\sim}A \nvdash B \tag{*}$$

---

[1]Cf. [5, 30–33].

[2]Here, (Abs) and (Ext) are $\exists y \forall x(x \in y \leftrightarrow B)$ and $\forall x(x \in z \leftrightarrow x \in y) \rightarrow z = y$ respectively where B is any formula which does not contain y free, and $\rightarrow$ and $\leftrightarrow$ are suitable conditional and biconditional.

where $\sim$ is a unary operation intended to be a negation and $\vdash$ is a logical consequence relation. Note here though that paraconsistentists even disagree on the understanding of negation. For example, Priest and Routley, in [34], famously argued against the systems of paraconsistent logic developed by Newton da Costa on the ground that the unary operation satisfying the above condition is not a negation.

Based on these, the aims of the paper are twofold. First, I present an account of negation based on the definition of paraconsistency given by Stanisław Jaśkowski, and briefly examine some of the existing systems of paraconsistent logic (Sect. 2). Second, I introduce a notion for a special kind of paraconsistent logic, called *dialetheic* logic (Sect. 3.1), and present an expansion of the Logic of Paradox (**LP** hereafter) of Priest that is dialetheic and suitable as the underlying logic for dialetheic theories based on the suggestion of Goodship. Some basics, including semantics and proof theory, are presented in Sect. 3.2, and this is followed by the soundness and completeness results in Sect. 3.3. I will then prove some of the distinctive features of the new system in Sect. 3.4, and consider two natural variants of the new system in Sect. 3.5.

2 Negation: Towards a Desideratum for Paraconsistent Logic

Negation[3] is the central connective for paraconsistency since it is the only connective used in stating the only criteria (*) for logics to be paraconsistent. In this section, we first revisit the definitions given by modern founders of paraconsistency, namely Stanisław Jaśkowski and Newton da Costa, and observe the difference in their understanding of negation. We then turn to present an account of negation which nicely realizes the idea of Jaśkowski. These will be followed by some observations on the existing systems of paraconsistent logic. Note here that the following observations are semantic, and more proof theoretic investigations will be kept for another occasion. Note also that the discussion in this section is far from being conclusive, and is meant to be a basis for further discussion.

The following definitions are given by Jaśkowski and da Costa respectively. Both definitions clearly distinguish inconsistent systems from trivial systems.

Definition 1 (Jaśkowski, [17]) A deductive system S is called *inconsistent*, if its theses include two such which contradict one another, that is such that one is the negation of the other, e.g., A and $\sim A$. A system in which any meaningful formula is a thesis shall be termed *overfilled*.[4]

Definition 2 (da Costa, [11]) A formal system (deductive system, deductive theory, ...) S is said to be *inconsistent* if there is a formula A of S such that A

[3]For an up-to-date survey on negation, see [16]. Note also that the following discussion focuses on the sentential negation since this is the key notion in the criteria for paraconsistent logics.

[4]Cf. [17, p.38]. The notation of negation is adjusted.

and its negation, $\sim A$, are both theorems of this system. In the opposite case, S is called *consistent*. A deductive system S is said to be *trivial* if all its formulas are theorems. If there is at least one unprovable formula in S, it is called *non-trivial*.[5]

At first glance, one may not find differences between the above two definitions. However, in my view, Jaśkowski's definition is slightly more informative than da Costa's definition, since Jaśkowski seems to be aiming at an understanding of negation through contradictory pairs of sentences. The details on contradictory pairs are not spelled out by Jaśkowski himself, so we need to fill in the gap. But how? There are many options, but one of the most popular definitions goes as follows.

Definition 3 (*C-Contradictories*) Let A and B be sentences. Then,

- A and B are *C-contraries* iff whenever one of them is untrue, the other is true;
- A and B are *C-subcontraries* iff whenever one of them is true, the other is untrue;
- A and B are *C-contradictories* iff they are C-contraries and C-subcontraries.

Another definition of contradictory pairs, which is probably less popular, goes as follows.

Definition 4 (*P-Contradictories*) Let A and B be sentences. Then,

- A and B are *P-contraries* iff whenever one of them is false, the other is true;
- A and B are *P-subcontraries* iff whenever one of them is true, the other is false;
- A and B are *P-contradictories* iff they are P-contraries and P-subcontraries.

Needless to say, C-contradictories and P-contradictories coincide if untruth and falsity are identified. But this does not have to be the case in general. This can be made clear with the help of formal language.

Let $\mathcal{L}$ be a propositional language that consists of a finite set of propositional connectives and a countable set Prop of propositional parameters. We assume that at least two unary operations $\neg_1$ and $\neg_2$ are included in the set of propositional language. We may of course include more connectives, but that is not necessary for the present purpose. Then, an interpretation for the language $\mathcal{L}$ is a relation, r, between propositional parameters and the values 1 and 0. More precisely, $r \subseteq \mathsf{Prop} \times \{1, 0\}$.

Once we have a formal device, it is easy to clarify the difference between untruth and falsity. Indeed, let p be a propositional parameter. Then p being untrue is represented as it is not the case that $pr1$ whereas p being false is represented as $pr0$. And since r is a relation, not a function, it is not necessarily the case that $pr0$ iff it is not the case that $pr1$.

Now, what Jaśkowski is suggesting is to capture the notion of negation through contradictory pairs, and if we make use of the formal language, the above definitions of C- and P-contradictories become as follows:

- A and B are C-contradictories iff (($Ar1$ iff not $Br1$) and (not $Ar1$ iff $Br1$))
- A and B are P-contradictories iff (($Ar1$ iff $Br0$) and ($Ar0$ iff $Br1$))

[5] Cf. [11, p.497]. The notation of negation is adjusted.

If we assume that the biconditional contraposes and that the double negation can be introduced and eliminated in the metalanguage, then the first condition will be simplified, and we obtain the following conditions.

- A and B are C-contradictories iff ($Ar1$ iff not $Br1$)
- A and B are P-contradictories iff (($Ar1$ iff $Br0$) and ($Ar0$ iff $Br1$))

These conditions suggest how to interpret unary operations intended to be negations. Indeed, let the truth condition for $\neg_1$, and the truth and falsity conditions for $\neg_2$ as follows.

- $\neg_1 Ar1$ iff not $Ar1$
- $\neg_2 Ar1$ iff $Ar0$
- $\neg_2 Ar0$ iff $Ar1$

Defined in this way, we obtain that A and $\neg_1 A$ are C-contradictories and that A and $\neg_2 A$ are P-contradictories. In other words, the above two negations seem to realize the idea of Jaśkowski.

Remark 5 In the above observation, we did not specify the falsity condition for $\neg_1$. The lack of falsity condition for $\neg_1$ implies that, in general, there are several operations that will form C-contradictories. See Remark 13 for such an example in an expansion of **LP**. Moreover, if we assume that 'not' in our metalanguage is both exclusive and exhaustive, and that semantic consequence relation is defined in the usual manner, then $\neg_1$ behaves exactly the same as the negation in classical logic.

Remark 6 One might argue against $\neg_1$ as a paraconsistent negation if we assume that 'not' in our metalanguage is both exclusive and exhaustive since it seems to imply the explosion principle immediately. This is true if we define the semantic consequence relation in the usual manner. However, there are some paraconsistent negations obtained with a relativized version of the truth condition for $\neg_1$ as follows.

- $\neg_1 Ar1$ at a world/state w iff not $Ar1$ at a world/state w

For example, Jaśkowski's discussive (or discursive) logic **D2** (cf. [17, 18]) is one such example. Indeed, if one follows the Kripke semantics for discussive logic as presented by Janusz Ciuciura in [10, Sect. 2], then one can observe that paraconsistency of **D2** is obtained by an unusual definition of semantic consequence relation which reflects the idea of Jaśkowski who originally defined **D2** through translation into modal logic **S5**.

Remark 7 The truth and falsity conditions for $\neg_2$ reflect a very simple idea of negation as flip-flopping the truth and falsity. And it deserves noting that such a simple account of negation is connected to a version of contradictory pairs, as we observed above.

Since we wish to keep the usual definition for the semantic consequence relation when the truth and falsity conditions are relativized, we focus on $\neg_2$ in the rest of this section.

Proposition 1 *The following facts hold for* $\neg_2$ *where the semantic consequence relation* $\models$ *is defined in terms of the preservation of truth.*

- $\neg_2\neg_2 A \models A$ *and* $A \models \neg_2\neg_2 A$ *for any* A*;*
- *For some* A *and* B*,* $A \models B$ *does* not *imply that* $\neg_2 B \models \neg_2 A$*;*
- *For some* A *and* B*,* $A \not\models \neg_2 A$ *and* $\neg_2 B \not\models B$*.*

The first shows that the validity of the laws of double negation introduction and elimination is a necessary condition for an unary operation to be negation satisfying the truth and falsity conditions for $\neg_2$. The second shows that contraposition, which is necessarily valid in some accounts of negation, such as those based on (in)compatibility semantics, is not necessarily valid for $\neg_2$. The last condition has been claimed by Wolfgang Lenzen [20] and João Marcos [21] to be a necessary condition on negation, and that is met by $\neg_2$.

Now, having an account of negation, the next question to ask is which unary operations in nonclassical logics are negation satisfying the truth and falsity conditions for $\neg_2$. The basic logics that are equipped with negation in the above sense include classical logic, **LP**, Kleene's strong three valued logic (**K3** hereafter), and the four-valued logic of Belnap and Dunn (**BD** hereafter). Indeed, the semantic condition for classical negation is obtained once we assume that truth and falsity are both exclusive and exhaustive. Moreover, if one assumes only one of exclusivity or exhaustivity, then we obtain the semantics for negations of **K3** and **LP** respectively. And finally, if we leave open the relation between truth and falsity, we obtain the semantics for the negation of **BD**. These facts imply that any of the expansions of the above systems contain negation. For example, modal logics that expand classical logic, **CLuNs** (cf. [4]) and **LFI1** (cf. [9]) that expand **LP**, Nelson logic **N3** (cf. [19, 40]) that expands **K3**, and relevant logics (à la American plan) that expand **BD** all contain negation in the above sense.[6]

Needless to say, there are some paraconsistent "negations" that are not counted as negation in the above sense. These include negation in systems such as many of the Logics of Formal Inconsistency (LFIs, cf. [7, 8]), including the base system **mbC**, and **CLuN** (cf. [3]). This observation follows by the fact that introduction of double negation fails in both systems. Note that these paraconsistent "negations" not being negation in our sense only means that those unary operations are meant to do something else than flip-flopping truth and falsity. I do *not* mean that those "negations" are incoherent or so. One may have an entirely different story to tell about those "negations", and that should be perfectly coherent as well.

[6]Note that we need the relativized truth and falsity conditions for modal logics, Nelson logics and relevant logics.

3 Dialetheic Logic

3.1 Preliminary Remarks

In the context of considering formal theories, one may view propositional logics as representing the most abstract structure of formal theories in the following sense. As an illustration, consider the classical Peano Arithmetic. Then we can first strip off all the axioms unique to Peano Arithmetic, and this leaves us with the classical predicate logic. However, one may take the further step to ignore the 'internal structure' of the sentences. This leaves us with the classical propositional logic. Then, in the case of classical arithmetic, we will have some formulas being provable which are represented by tautologies in propositional logic. But what is the characteristic feature of dialetheic theories? As one may expect, some formulas and their negations will both be provable.

In sum, if one agrees with the above way to look at logic, then it is natural to require logic to have sufficient expressive power so that a formula, representing a dialetheia, be definable at the level of propositional logic. And once this requirement is met, we may distinguish this special kind of paraconsistent logic from other paraconsistent logics.[7] I will refer to this special kind of paraconsistent logics as *dialetheic logic* for the obvious reason. With this remark in mind, most systems of paraconsistent logic deployed in developing dialetheic theories are *not* dialetheic. For example, **LP** is not dialetheic since it is a subsystem of classical logic. As another example, Zach Weber proves some interesting results in naive set theory based on a relevant logic in [43, 44], marking a big progress in dialetheic set theory. But again, the underlying logic is not dialetheic for the same reason.

Now, there are various approaches in developing dialetheic theories of truth and sets. One of the approaches is to follow the suggestion by Goodship in [15]. More specifically, Goodship pointed out the advantages of formulating dialetheic theories in terms of material biconditional defined in terms of paraconsistent negation. As for naive set theory based on **LP**, it is proved to be non-trivial by Greg Restall in [36], and this carries over even when classical negation is definable in the underlying logic such as **LFI1**, an expansion of **LP** (cf. [25]). However, **LFI1** is *not* dialetheic, since it can be regarded as a subsystem of classical logic. In particular, **LFI1** is not expressively full since the matrix of **LFI1** is not functionally complete,[8] where functional completeness is one of the common standards in measuring the expressive power of many-valued logics. But, one may motivate having fully expressive logic following the reasons that classical logicians have. And assuming the motivations being reasonable, I will present an expansion of **LP**, referred to as **dLP** (dialetheic **LP**), whose matrix is functionally complete. Note here that when it comes to functional

[7] One may of course have some strong arguments against such a view on logic, and if that is the case, then the above expressivity requirement will not be substantial.

[8] A simple way to see this is that 'classical' values are closed under the operations in **LFI1**, and thus the constant function mapping every argument to the intermediate value is not definable.

completeness, there are already some detailed studies in the literature (cf. [1, 2, 23, 35, 37]). The novelty of the observation to follow is in the fact that the expansion involves two different traditions in paraconsistency: Logics of Formal Inconsistency and connexive logic. Connexive logics are characterized as logics having the following formulas as theorems of the system.[9]

$$\sim(A \to \sim A) \tag{AT}$$

$$(A \to B) \to \sim(A \to \sim B) \tag{BT}$$

$$\sim(\sim A \to A) \tag{AT'}$$

$$(A \to \sim B) \to \sim(A \to B) \tag{BT'}$$

As one can see, none of the above formulas are provable in classical logic, and thus connexive logics are highly nonclassical logics. For an up-to-date survey on the topic, see [42].[10]

3.2 Basics

The language $\mathcal{L}_\circ$ consists of the following vocabulary: a set $\{\sim, \circ, \wedge, \vee, \to\}$ of propositional connectives, the universal and particular quantifiers $\forall$ and $\exists$, a countable set $\{x_0, x_1, \dots\}$ of variables, a countable set $\{c_0, c_1, \dots\}$ of constant symbols, and a countable set $\{P_0, P_1, \dots\}$ of predicate symbols, where we associate each predicate P_k with a fixed finite arity. We regard 0-ary predicate symbols as propositional letters. We define the set of formulas in $\mathcal{L}_\circ$ as follows:

$$A ::= P(t_1, \dots, t_n) \mid \sim A \mid \circ A \mid A \wedge B \mid A \vee B \mid A \to B \mid \forall x A \mid \exists x A,$$

where t_i is a *term*, namely a variable or a constant symbol. We say that a formula is *propositional* if it is constructed from propositional letters (i.e., 0-ary predicate symbols) by using the propositional connectives. We define the notions of *free* and *bound* variable, and *sentence* as usual. We write $A_x(t)$ to mean the result of substituting all the occurrences of free variable x in A by the term t, renaming the bound variables, if necessary, to avoid variable-clashes. We denote sets of formulas by Γ, Σ, etc.

Now I introduce the semantics.

---

[9]Note that connexive logics are not necessarily paraconsistent in general. But the idea imported in expanding **LP** relies on a kind of connexive logics that are also paraconsistent, and this is why I counted connexive logic as a tradition in paraconsistency.

[10]See also [22] for a survey by Storrs McCall, one of the modern founders of connexive logics.

Definition 8 An *interpretation* $\mathcal{I}$ is a pair $\langle D, v\rangle$ where D is a non-empty set D and we assign $v(c) \in D$ to each constant c, assign both the *extension* $v^{+}(P) \subseteq D^{n}$ and the *anti-extension* $v^{-}(P) \subseteq D^{n}$ to each n-ary predicate symbol P where $v^{+}(P) \cup v^{-}(P) = D^{n}$. Given any interpretation $\langle D, v\rangle$, we can define **dLP**-valuation $\overline{v}$ for all the sentences of $\mathcal{L}_{\circ}$ expanded by $\{k_d : d \in D\}$ inductively as follows: as for the atomic *sentences*,

$$1 \in \overline{v}(P(t_1, ..., t_n)) \text{ iff } \langle v(t_1), \ldots, v(t_n)\rangle \in v^{+}(P),$$
$$0 \in \overline{v}(P(t_1, ..., t_n)) \text{ iff } \langle v(t_1), \ldots, v(t_n)\rangle \in v^{-}(P).$$

The rest of the clauses are as follows:

$1 \in \overline{v}(\sim A)$	iff $0 \in \overline{v}(A)$,	$0 \in \overline{v}(\sim A)$	iff $1 \in \overline{v}(A)$,
$1 \in \overline{v}(\circ A)$	iff $1 \notin \overline{v}(A)$ or $0 \notin \overline{v}(A)$,	$0 \in \overline{v}(\circ A)$	iff $1 \in \overline{v}(A)$ and $0 \in \overline{v}(A)$,
$1 \in \overline{v}(A \wedge B)$	iff $1 \in \overline{v}(A)$ and $1 \in \overline{v}(B)$,	$0 \in \overline{v}(A \wedge B)$	iff $0 \in \overline{v}(A)$ or $0 \in \overline{v}(B)$,
$1 \in \overline{v}(A \vee B)$	iff $1 \in \overline{v}(A)$ or $1 \in \overline{v}(B)$,	$0 \in \overline{v}(A \vee B)$	iff $0 \in \overline{v}(A)$ and $0 \in \overline{v}(B)$,
$1 \in \overline{v}(A \rightarrow B)$	iff $1 \notin \overline{v}(A)$ or $1 \in \overline{v}(B)$,	$0 \in \overline{v}(A \rightarrow B)$	iff $1 \notin \overline{v}(A)$ or $0 \in \overline{v}(B)$,
$1 \in \overline{v}(\forall x A)$	iff $1 \in \overline{v}(A_x(k_d))$, for all $d \in D$,	$0 \in \overline{v}(\forall x A)$	iff $0 \in \overline{v}(A_x(k_d))$, for some $d \in D$,
$1 \in \overline{v}(\exists x A)$	iff $1 \in \overline{v}(A_x(k_d))$, for some $d \in D$,	$0 \in \overline{v}(\exists x A)$	iff $0 \in \overline{v}(A_x(k_d))$, for all $d \in D$.

Finally, let $\Gamma \cup \{A\}$ be any set of sentences. Then, A is a **dLP**-*semantic consequence* from Γ ($\Gamma \models_{\mathbf{dLP}} A$) iff for every interpretation $\mathcal{I} = \langle D, v\rangle$ and for every **dLP**-valuation $\overline{v}$, $1 \in \overline{v}(A)$ if $1 \in \overline{v}(B)$ for all $B \in \Gamma$.

Remark 9 First, note that the truth and falsity conditions for $\sim$ are those for $\neg_2$ in the previous section. Second, the falsity condition for the conditional being quite different from the more familiar clause "$1 \in \overline{v}(A)$ and $0 \in \overline{v}(B)$". This is the key to obtain the connexive conditional introduced by Heinrich Wansing in [41]. Moreover, the truth tables for the propositional connectives become as follows[11]:

A	$\sim A$	$\circ A$
t	**f**	**t**
b	**b**	**f**
f	**t**	**t**

$A \wedge B$	**t**	**b**	**f**
t	**t**	**b**	**f**
b	**b**	**b**	**f**
f	**f**	**f**	**f**

$A \vee B$	**t**	**b**	**f**
t	**t**	**t**	**t**
b	**t**	**b**	**b**
f	**t**	**b**	**f**

$A \rightarrow B$	**t**	**b**	**f**
t	**t**	**b**	**f**
b	**t**	**b**	**f**
f	**b**	**b**	**b**

Note here that designated values are **t** and **b**. The only difference from the truth tables for **LFI1** lies in the truth table for $\rightarrow$, now having the value **b**, not **t**, when A is assigned the value **f**. This implies that the propagation of consistency over the conditional fails in **dLP**. Indeed, the formula $(\circ A \wedge \circ B) \rightarrow \circ(A \rightarrow B)$ takes the value **f** when A and B are both assigned the value **f**.[12]

[11] Note here that if expansions of **BD** is concerned, then not only that we obtain truth tables from truth and falsity conditions of relational semantics, but we can also go the other way around *mechanically*, namely to obtain truth and falsity conditions of relational semantics out of given any truth tables. For the details, see [27].

[12] For an examination of the propagation of consistency in LFIs, see [29].

Remark 10 There is a closely related system in the literature developed by John Cantwell in [6]. More precisely, he takes the propositional language $\mathcal{L}$ expanded by $\bot$ ($\mathcal{L}_\bot$ hereafter), not $\circ$, where $\bot$ is always assigned the value **f**, and other connectives are exactly as in **dLP**. Then, a natural question to ask is the relation between Cantwell's logic **CN** and **dLP**. The answer is that **dLP** is strictly stronger than **CN**. More specifically, $\circ$ is *not* definable in **CN**. (The details are spelled out in the appendix.) Recall here that if the conditional is taken to be the one in **LFI1**, then expanding the language by $\bot$ and $\circ$ have the same effect which shows the equivalence of the two systems **LFI1** and **CLuNs**.

Note also that Grigory Olkhovikov introduced a three-valued logic equivalent to **dLP** in [24].[13] One of the differences lies in the language. More specifically, Olkhovikov's system includes a unary operation L instead of $\circ$, and L is characterized by the following truth table.

A	$\mathsf{L}A$
t	**t**
b	**f**
f	**f**

It is easy to observe the equivalence of the two systems. Indeed, $\circ A$ is definable in Olkhovikov's system as $\mathsf{L}A \vee \mathsf{L}{\sim}A$, and $\mathsf{L}A$ is definable in **dLP** as $\circ A \wedge A$.

Now I turn to the proof theory.

Definition 11 The system **dLP** consists of the following axioms and rules of inference where $A \leftrightarrow B =_{\text{def.}} (A \to B) \wedge (B \to A)$.

$$A \to (B \to A) \tag{A1}$$

$$(A \to (B \to C)) \to ((A \to B) \to (A \to C)) \tag{A2}$$

$$((A \to B) \to A) \to A \tag{A3}$$

$$A \to (A \vee B) \tag{A4}$$

$$B \to (A \vee B) \tag{A5}$$

$$(A \to C) \to ((B \to C) \to ((A \vee B) \to C)) \tag{A6}$$

$$(A \wedge B) \to A \tag{A7}$$

$$(A \wedge B) \to B \tag{A8}$$

[13] This was discovered after the first submission. I would like to thank Heinrich Wansing who informed me of Olkhovikov's system, and Grigory Olkhovikov who sent me his paper and translated some of the results during a discussion.

$$(C \to A) \to ((C \to B) \to (C \to (A \wedge B))) \tag{A9}$$

$$A \vee {\sim}A \tag{A10}$$

$$\circ A \to ((A \wedge {\sim}A) \to B) \tag{A11}$$

$${\sim}{\sim}A \leftrightarrow A \tag{A12}$$

$${\sim}(A \wedge B) \leftrightarrow ({\sim}A \vee {\sim}B) \tag{A13}$$

$${\sim}(A \vee B) \leftrightarrow ({\sim}A \wedge {\sim}B) \tag{A14}$$

$${\sim}(A \to B) \leftrightarrow (A \to {\sim}B) \tag{A15}$$

$${\sim}(\circ A) \leftrightarrow (A \wedge {\sim}A) \tag{A16}$$

$$\forall x A(x) \to A(a) \text{ where } a \text{ is a term.} \tag{A17}$$

$$A(a) \to \exists x A(x) \text{ where } a \text{ is a term.} \tag{A18}$$

$${\sim}\forall x A(x) \leftrightarrow \exists x {\sim}A(x) \tag{A19}$$

$${\sim}\exists x A(x) \leftrightarrow \forall x {\sim}A(x) \tag{A20}$$

$$\frac{A \quad A \to B}{B} \tag{MP}$$

$$\frac{B \to A(x)}{B \to \forall x A(x)} \tag{$\forall$-intro}$$

$$\frac{A(x) \to B}{\exists x A(x) \to B} \tag{$\exists$-intro}$$

where x is not free in B of ($\forall$-intro) and ($\exists$-intro).

Finally, if Σ is a set of formulas and A is a formula, then $\Gamma \vdash_{\mathbf{dLP}} A$ iff there is a sequence of formulas $B_1, \ldots, B_n, A, n \geq 0$, such that every formula in the sequence $B_1, \ldots, B_n, A$ either (i) belongs to Γ; (ii) is an axiom of **dLP**; (iii) is obtained by (MP), ($\forall$-intro) or ($\exists$-intro) from formulas preceding it in sequence.

Remark 12 The distinctive axiom here is (A15) that gives us the falsity condition of the conditional. The more familiar form is ${\sim}(A \to B) \leftrightarrow (A \wedge {\sim}B)$, and if we replace (A15) by the familiar one, then we obtain the axiomatization for the three-valued logic known as **LFI1** (cf. [9, 28]), which is equivalent to $\mathbf{J}_3$ of d'Ottaviano and da Costa (cf. [13, 14]), and to **CLuNs** (cf. [4]).

Remark 13 We may define two unary operations: $\neg_3 A := {\sim}A \wedge \circ A$ and $\neg_4 A := A \to \bot$ where $\bot =_{\text{def.}} \circ B \wedge B \wedge {\sim}B$ for some B. Semantically, these two operations satisfy the truth condition for $\neg_1$ discussed in the previous section. Thus we may conclude that these two negations are classical negations. However, these two negations do not coincide, unlike in **LFI1**, due to (A15). More specifically, for $\neg_4$, we have that ${\sim}\neg_4 A$ is derivable for any A since ${\sim}\neg_4 A$ is ${\sim}(A \to \bot)$ by the definition of $\neg_4$, and this is equivalent to $A \to {\sim}\bot$ in view of (A15), and this is provable in **dLP**. This does not hold for $\neg_3$, however. Semantically speaking, the falsity conditions for $\neg_3$ and $\neg_4$ are as follows.

- $\neg_3 Ar0$ iff $Ar1$,
- $\neg_4 Ar0$.

That is, $\neg_3 A$ is false iff A is true, and $\neg_4 A$ is always false. This is an example in which two different operations both form C-contradictories (cf. Remark 5). In what follows, we will refer to $\neg_3$ as $\neg$. Note finally that we do have the biconditional $\neg A \leftrightarrow (A \to \bot)$ as a *provable* formula, as observed in the following proposition.

Proposition 2 *The following formulas are provable in* **dLP**.

$$\begin{align}
&{\sim}(A \to {\sim}A) \tag{1}\\
&{\sim}({\sim}A \to A) \tag{2}\\
&(A \to B) \to {\sim}(A \to {\sim}B) \tag{3}\\
&(A \to {\sim}B) \to {\sim}(A \to B) \tag{4}\\
&(A \wedge {\sim}B) \to {\sim}(A \to B) \tag{5}
\end{align}$$

$$\begin{align}
&\circ A \leftrightarrow \neg(A \wedge {\sim}A) \tag{6}\\
&\neg A \leftrightarrow (A \to \bot) \tag{7}\\
&(A \wedge \neg A) \to B \tag{8}\\
&A \vee \circ A \tag{9}\\
&A \vee (A \to B) \tag{10}
\end{align}$$

Proof Equations (1) and (2) are obtained by applying (A15) to $(A \to {\sim}{\sim}A)$ and $({\sim}A \to {\sim}A)$ respectively. As for (3), we again use (A15), and (4) is the right-to-left direction of (A15). Equation (5) immediately follows in view of (4). As for (6), the left-to-right direction is immediate in view of (A11). For the other way around, we make use of (A16) and (A10). As for (7), the left-to-right direction is immediate again in view of (A11). For the other way around, we need to prove $(A \to \bot) \to {\sim}A$ and $(A \to \bot) \to \circ A$. But these are easy to prove in view of (A10) and (A16) respectively. Equation (8) follows immediately in view of (7). For (9), we make use of (A10) and (A16). Finally, for (10), we first obtain $A \vee \neg A$ by (A10) and (9), and the desired result follows by (7). □

Remark 14 Equations (1)–(4) show that **dLP** is a connexive logic, as expected. Moreover, (5) shows that $A \wedge {\sim}B$ is sufficient for the conditional $A \rightarrow B$ to be false, but not necessary. Indeed, if we assign **f** and **t** for A and B respectively then the formula ${\sim}(A \rightarrow B) \rightarrow (A \wedge {\sim}B)$ takes the value **f**.

3.3 Soundness and Completeness

We now turn to the soundness and completeness results. The soundness is routine as usual.

Proposition 3 (Soundness) *Given any set of sentences* $\Gamma \cup \{A\}$, *if* $\Gamma \vdash_{\mathbf{dLP}} A$ *then* $\Gamma \models_{\mathbf{dLP}} A$.

Proof By induction on the derivation $\Gamma \vdash_{\mathbf{dLP}} A$, as usual. □

For the completeness, we need the following notions.

Definition 15 Let Σ be a set of formulas. Then,

- Σ is a *theory* if it is closed under $\vdash_{\mathbf{dLP}}$, i.e., if $\Sigma \vdash_{\mathbf{dLP}} A$ then $A \in \Sigma$ for any formula A;
- Σ is *prime* if $A \vee B \in \Sigma$ implies that $A \in \Sigma$ or $B \in \Sigma$ for any A and B;
- Σ is *non-trivial* if for some formula A, $A \notin \Sigma$;
- Σ is *saturated* if the following holds:
  - $\forall x A \in \Gamma$ iff $A_x(c) \in \Gamma$ for any constant c, and
  - $\exists x A \in \Gamma$ iff $A_x(c) \in \Gamma$ for some constant c.

The rest of the proof is quite standard.

Lemma 1 *If* Γ *is a prime theory, then* $A \rightarrow B \in \Gamma$ *iff* ($A \notin \Gamma$ *or* $B \in \Gamma$).

Proof For the left-to-right direction, suppose that $A \rightarrow B \in \Gamma$ and that $A \in \Gamma$ and $B \notin \Gamma$. Then by the first two conditions and (MP), we obtain $\Gamma \vdash_{\mathbf{dLP}} B$, and since Γ is a theory, we have $B \in \Gamma$, but this contradicts to the third condition. For the other direction, it suffices to prove that (i) if $A \notin \Gamma$ then $A \rightarrow B \in \Gamma$, and (ii) if $B \in \Gamma$ then $A \rightarrow B \in \Gamma$. For (i), assume that $A \notin \Gamma$ and $A \rightarrow B \notin \Gamma$. Then since Γ is prime, we obtain $A \vee (A \rightarrow B) \notin \Gamma$. Moreover, since Γ is a theory, we obtain $\Gamma \nvdash_{\mathbf{dLP}} A \vee (A \rightarrow B)$, but this is a contradiction in view of (10). For (ii), if $B \in \Gamma$, then $\Gamma \vdash_{\mathbf{dLP}} B$, and by (A1) and (MP), we obtain $\Gamma \vdash_{\mathbf{dLP}} A \rightarrow B$. Since Γ is a theory, we obtain the desired result. This completes the proof. □

Lemma 2 *If* Γ *is a non-trivial prime theory, then* $\circ A \in \Gamma$ *iff* ($A \notin \Gamma$ *or* ${\sim}A \notin \Gamma$).

Proof For the left-to-right direction, suppose that $\circ A \in \Gamma$ and that $A \in \Gamma$ and $\sim A \in \Gamma$. Then, in view of (A11), we obtain $\Gamma \vdash_{\mathbf{dLP}} B$ for any B, and thus $B \in \Gamma$ since Γ is a theory. But this contradicts the assumption that Γ is non-trivial. For the other direction, it suffices to prove that (i) if $A \notin \Gamma$ then $\circ A \in \Gamma$, and (ii) if $\sim A \notin \Gamma$ then $\circ A \in \Gamma$. For (i), assume that $A \notin \Gamma$ and $\circ A \notin \Gamma$. Then since Γ is prime, we obtain $A \vee \circ A \notin \Gamma$. Moreover, since Γ is a theory, we obtain $\Gamma \nvdash_{\mathbf{dLP}} A \vee \circ A$, but this is a contradiction in view of (9). The proof for (ii) is similar to (i), and this completes the proof. □

Lemma 3 *Let* $\Gamma \cup \{A\}$ *be any set of sentences. If* $\Gamma \nvdash_{\mathbf{dLP}} A$*, then by adding countably new constant symbols, we can extend* $\langle \Gamma, \{A\} \rangle$ *to* $\langle \Gamma^+, \Pi^+ \rangle$ *such that* $\Gamma \subseteq \Gamma^+$, $A \in \Pi^+$, $\Gamma^+ \nvdash_{\mathbf{dLP}} \Pi^+$, *either* $B \in \Gamma^+$ *or* $B \in \Pi^+$ *holds for all* B, *and* Γ^+ *is a prime and saturated theory.*

Proof Let us expand our language with a countable set $E := \{e_n : n \in \omega\}$ of fresh constant symbols. Let $(A_n)_{n \geq 1}$ be an enumeration of all formulas in the expanded syntax. We inductively define the sequence $(\langle \Gamma_n, \Pi_n \rangle)_{n \in \omega}$ such that $\Gamma_n \nvdash_{\mathbf{dLP}} \Pi_n$ as follows:

- $\Gamma_0 := \Gamma$ and $\Pi_0 := \{A\}$.
- Suppose that we have constructed $\langle \Gamma_{n-1}, \Pi_{n-1} \rangle$ such that $\Gamma_{n-1} \nvdash_{\mathbf{dLP}} \Pi_{n-1}$. We have the following two cases:
  - if $\Gamma_{n-1} \cup \{A_n\} \nvdash_{\mathbf{dLP}} \Pi_{n-1}$, then we split the case depending on the form of A_n:
    If $A_n = \exists x B$, we define $\Gamma_n := \Gamma_{n-1} \cup \{A_n, B_x(e)\}$ and $\Pi_n := \Pi_{n-1}$, where e is the first constant in the enumeration of E such that it is fresh in Γ_{n-1}, Π_{n-1} and A_n.
    Otherwise, $\Gamma_n := \Gamma_{n-1} \cup \{A_n\}$ and $\Pi_n := \Pi_{n-1}$.
  - If $\Gamma_{n-1} \cup \{A_n\} \vdash_{\mathbf{dLP}} \Pi_{n-1}$, then we again split the case depending on the form of A_n:
    If $A_n = \forall x B$, we define $\Gamma_n := \Gamma_{n-1}$ and $\Pi_n := \Pi_{n-1} \cup \{A_n, B_x(e)\}$, where e is the first constant in the enumeration of E such that it is fresh in Γ_{n-1}, Π_{n-1} and A_n.
    Otherwise, we put $\Gamma_n := \Gamma_{n-1}$ and $\Pi_n := \Pi_{n-1} \cup \{A_n\}$.

In both cases, it is easy to see that $\Gamma_n \nvdash_{\mathbf{dLP}} \Pi_n$.

We define the 'limit' of the sequence $(\langle \Gamma_n, \Pi_n \rangle)_{n \in \omega}$ as $\Gamma^+ := \bigcup_{n \in \omega} \Gamma_n$ and $\Pi^+ := \bigcup_{n \in \omega} \Pi_n$. It is clear that $\Gamma^+ \nvdash_{\mathbf{dLP}} \Pi^+$. By construction, $A \in \Gamma^+$ or $A \in \Pi^+$ for any A. Moreover, Γ^+ is a prime and saturated theory. Here we only show the saturation requirement for $\forall$ of Γ^+: $\forall x A \in \Gamma^+$ iff $A_x(c) \in \Gamma^+$ for any constant c. The left-to-right direction is not difficult to show (if we assume that we have established Γ^+ is a theory). As for the converse, we show the contrapositive. Assume that $\forall x A \notin \Gamma^+$. Since $\forall x A \in \Gamma^+$ or $\forall x A \in \Pi^+$, we have $\forall x A \in \Pi^+$, which implies $A_x(e) \in \Pi^+$ for some e by construction. Then, we obtain $A_x(e) \notin \Gamma^+$ as follows. Suppose for reductio that $A_x(e) \in \Gamma^+$. Then, it follows from $A_x(e) \in \Pi^+$ that $\Gamma^+ \vdash_{\mathbf{dLP}} \Pi^+$, a contradiction. □

Now we are ready to prove the completeness result.

Theorem 1 (Completeness) *Given any set of sentences* $\Gamma \cup \{A\}$, *we have* $\Gamma \models_{\mathbf{dLP}} A$ *iff* $\Gamma \vdash_{\mathbf{dLP}} A$.

Proof Since we have already observed the soundness, we here prove the completeness part. And to this end, we prove the contrapositive. Assume that $\Gamma \nvdash_{\mathbf{dLP}} A$. Then by Lemma 3, there is a prime and saturated theory Γ^+ such that $\Gamma \subseteq \Gamma^+$ and $\Gamma^+ \nvdash_{\mathbf{dLP}} A$. Now, define an interpretation $\mathcal{I}_{\Gamma^+} = \langle D, v \rangle$ as follows: $D = \{c : c \text{ is a constant symbol}\}$ and, for any n-ary predicate symbol P:

$$v^+(P) := \{\langle t_1, \ldots, t_n \rangle : P(t_1, \ldots, t_n) \in \Gamma^+\},$$
$$v^-(P) := \{\langle t_1, \ldots, t_n \rangle : {\sim}P(t_1, \ldots, t_n) \in \Gamma^+\},$$

and, for any constant symbol c, $v(c) = c$. Then, for any sentence A, the following holds.

$$1 \in \overline{v}(A) \text{ iff } A \in \Gamma^+,$$
$$0 \in \overline{v}(A) \text{ iff } {\sim}A \in \Gamma^+.$$

This can be proved by induction on A. We will here only check the cases in which A is of the form $\circ B$ and $B \to C$. For the positive case for the consistency operator,

$$\begin{aligned} 1 \in \overline{v}(\circ B) &\text{ iff } 1 \notin \overline{v}(B) \text{ or } 0 \notin \overline{v}(B) && \\ &\text{ iff } B \notin \Gamma^+ \text{ or } {\sim}B \notin \Gamma^+ && \text{IH} \\ &\text{ iff } \circ B \in \Gamma^+. && \text{Lemma 2} \end{aligned}$$

For the negative case for the consistency operator,

$$\begin{aligned} 0 \in \overline{v}(\circ B) &\text{ iff } 1 \in \overline{v}(B) \text{ and } 0 \in \overline{v}(B) && \\ &\text{ iff } B \in \Gamma^+ \text{ and } {\sim}B \in \Gamma^+ && \text{IH} \\ &\text{ iff } B \wedge {\sim}B \in \Gamma^+ && \Gamma^+ : \text{theory} \\ &\text{ iff } {\sim}\circ B \in \Gamma^+. && \text{(A16)} \end{aligned}$$

For the positive case for the conditional,

$$\begin{aligned} 1 \in \overline{v}(B \to C) &\text{ iff } 1 \notin \overline{v}(B) \text{ or } 1 \in \overline{v}(C) && \\ &\text{ iff } B \notin \Gamma^+ \text{ or } C \in \Gamma^+ && \text{IH} \\ &\text{ iff } B \to C \in \Gamma^+. && \text{Lemma 1} \end{aligned}$$

For the negative case for the conditional,

$$\begin{aligned} 0 \in \overline{v}(B \to C) &\text{ iff } 1 \notin \overline{v}(B) \text{ or } 0 \in \overline{v}(C) && \\ &\text{ iff } B \notin \Gamma^+ \text{ or } {\sim}C \in \Gamma^+ && \text{IH} \\ &\text{ iff } B \to {\sim}C \in \Gamma^+ && \text{Lemma 1} \\ &\text{ iff } {\sim}(B \to C) \in \Gamma^+. && \text{(A15)} \end{aligned}$$

Therefore we obtain the desired result since we have that $1 \in \overline{v}(C)$ for any $C \in \Gamma^+$ and that $1 \notin \overline{v}(A)$ (since $\Gamma^+ \nvdash_{\mathbf{dLP}} A$, i.e. $A \notin \Gamma^+$), that is, $\Gamma \not\models_{\mathbf{dLP}} A$. This completes the proof. □

3.4 *Some Distinctive Features of* dLP

I now turn to observe three distinctive features of the propositional fragment of **dLP**: inconsistency, definitional completeness, and Post completeness. First, we observe the inconsistency of **dLP**.

Proposition 4 $\vdash_{\mathbf{dLP}} (A \wedge \neg A) \rightarrow B$ *and* $\vdash_{\mathbf{dLP}} {\sim}((A \wedge \neg A) \rightarrow B)$. *Thus,* **dLP** *itself is inconsistent.*

Proof The first result is (8). The second result is also immediate in view of the first result and (A15). □

Remark 16 Another example that observes the inconsistency of **dLP** is the following (cf. [42]):

$$\vdash_{\mathbf{dLP}} (A \wedge {\sim}A) \rightarrow ({\sim}A \vee A) \text{ and } \vdash_{\mathbf{dLP}} {\sim}((A \wedge {\sim}A) \rightarrow ({\sim}A \vee A)).$$

An even simpler example is the following:

$$\vdash_{\mathbf{dLP}} (A \wedge {\sim}A) \rightarrow {\sim}A \text{ and } \vdash_{\mathbf{dLP}} {\sim}((A \wedge {\sim}A) \rightarrow {\sim}A).$$

This can already be derived by combing elimination of conjunction (A7), (A8), Boethius' thesis (BT), and Modus Ponens (MP). Thus we may safely conclude that connexive logics are very "close" to paraconsistent logics even though connexive logics are not paraconsistent in general.

Second, we observe the definitional completeness of **dLP**. To this end, we introduce some notions.

Definition 17 (*Functional completeness*) A matrix $\langle \mathfrak{A}, B \rangle$ where $\mathfrak{A} = \langle A, f_1, \ldots, f_n \rangle$, is said to be *functionally complete* provided that every function $f : A^n \rightarrow A$ is definable by superpositions of the functions $f_1, \ldots, f_n$ alone.

Definition 18 (*Definitional completeness*) A logic **L** is *definitionally complete* if there exists a functionally complete matrix that is strongly adequate for L.

For the characterization of the functional completeness, the following theorem of Jerzy Słupecki is elegant and useful. In order to state the result, we need the following definition.

Definition 19 Let $\mathfrak{A}$ be an algebra, and f be a binary operation defined in $\mathcal{F}$. Then, f is *unary reducible* iff for some unary operation g definable in $\mathcal{F}$, $f(x, y) = g(x)$ for all $x, y \in \mathfrak{A}$ or $f(x, y) = g(y)$ for all $x, y \in \mathfrak{A}$. And f is *essentially binary* if f is not unary reducible.

Theorem 2 (Słupecki, [38]) $\mathfrak{A}$ $(\sharp\mathcal{V} \geq 3)$ *is functionally complete iff in* $\mathfrak{A}$

(i) all unary functions on $\mathcal{V}$ are definable, and
(ii) at least one surjective and essentially binary function on $\mathcal{V}$ is definable.

Based on this characterization by Słupecki, the desired result is obtained as follows.

Theorem 3 **dLP** *is definitionally complete.*

Proof With the help of Theorem 2 what we need to show is that **dLP** satisfies the above two conditions (i) and (ii) of Theorem 2. However, (ii) is already met by the presence of $\wedge$ (or $\vee$), so the remaining task is to show that all unary operations on $\{\mathbf{t}, \mathbf{b}, \mathbf{f}\}$ are definable. Now assume that the following unary operations are definable where $a \in \{\mathbf{t}, \mathbf{b}, \mathbf{f}\}$:

x	$\delta_{\mathbf{t}}(x)$	$\delta_{\mathbf{b}}(x)$	$\delta_{\mathbf{f}}(x)$	$\mathsf{C}_a(x)$
t	**t**	**f**	**f**	a
b	**f**	**t**	**f**	a
f	**f**	**f**	**t**	a

Then any unary operation $\varphi(x)$ can be defined as follows:

$$\varphi(x) = \bigvee_{a\in\{\mathbf{t},\mathbf{b},\mathbf{f}\}} (\mathsf{C}_{\varphi(a)}(x) \wedge \delta_a(x))$$

Finally, the definability of the above operations are easy. Indeed, $x \wedge \circ x$, $\sim \circ x$, and $\sim x \wedge \circ x$ define $\delta_{\mathbf{t}}(x)$, $\delta_{\mathbf{b}}(x)$, and $\delta_{\mathbf{f}}(x)$ respectively, and $\circ\circ x$, $(x\wedge\neg x)\rightarrow y$, and $\sim\circ\circ x$ define $\mathsf{C}_{\mathbf{t}}(x)$, $\mathsf{C}_{\mathbf{b}}(x)$, and $\mathsf{C}_{\mathbf{f}}(x)$ respectively. This completes the proof. □

Remark 20 In fact, the algebra without conjunction and disjunction is already functionally complete. The details are spelled out in the appendix.

Finally, the Post completeness of **dLP** is observed.

Definition 21 The logic **L** is *Post complete* iff for every formula A such that $\nvdash A$, extension of **L** by A becomes trivial, i.e. $\vdash_{\mathbf{L}\cup\{A\}} B$ for any B.

Theorem 4 (Tokarz, [39]) *Definitionally complete logics are Post complete.*

In view of Theorems 3 and 4, we obtain the following result.

Corollary 1 **dLP** *is Post complete.*

3.5 Two Variants of dLP

Once some basic results are established, it is also natural to consider some variants of **dLP**. Here we consider two of them. First, the logic **dLP** will be formulated in a different language, and second, the logic **BD**, instead of **LP**, will be taken as the base system.

So far, the consistency operator is taken to be the distinctive notion for the systems of paraconsistent logic in the tradition of da Costa. However, sometimes it has been also characterized as having classical negation definable in the logic.[14] Based on this, we introduce a language $\mathcal{L}_\neg$ having $\neg$ instead of $\circ$, and consider a variant of **dLP**, called **dLP**′, in $\mathcal{L}_\neg$. Semantically, **dLP**′-valuation is obtained by replacing the truth and falsity conditions for $\circ$ by the following conditions for $\neg$.

$$1 \in \overline{v}(\neg A) \text{ iff } 1 \notin \overline{v}(A),\ 0 \in \overline{v}(\neg A) \text{ iff } 1 \in \overline{v}(A),$$

The truth table for $\neg$ is as follows.

A	$\neg A$
t	**f**
b	**f**
f	**t**

Proof theoretically, **dLP**′ is obtained by first dropping (A11) and (A16), and second adding the following axioms for $\neg$:

$$A \to (\neg A \to B) \qquad (A\neg 1)$$

$$A \vee \neg A \qquad (A\neg 2)$$

$$\sim\neg A \leftrightarrow A \qquad (A\neg 3)$$

Then, the soundness and completeness results carry over from **dLP**. Moreover, **dLP**′, is essentially equivalent to **dLP**. Indeed, we have the following result.

Proposition 5 *$\neg A$ is definable in* **dLP** *by $\sim A \wedge \circ A$ (or $\sim\circ(A\to\sim\circ\circ A)$) if conjunction is not available), and $\circ A$ is definable in* **dLP**′ *by $\neg(A \wedge \sim A)$ (or $\neg\neg(\sim A\to\neg A)$) if conjunction is not available).*

Proof One may easily check through truth tables. □

I now turn to the four-valued case, namely the case in which **BD** is taken as the base logic. On the one hand, if we take the language $\mathcal{L}_\circ$, then we face the problem about the reading of $\circ$. One way is to stick to the reading of the connective as *consistency* operator. Another way is to read the connective as *classicality* operator. If we follow

[14] For some discussions on classical negation in expansions of Belnap-Dunn logic, see [12].

the latter reading, then it is proved in [26, Theorem 4] that classical negation is *not* definable in the expansion of **BD** by the classicality operator. Therefore, two cases starting with **LP** and **BD** are not completely parallel.

On the other hand, if we take the language $\mathcal{L}_{\neg}$, then we face the problem of choosing one of the classical negations out of 16 candidates (cf. [12]). If we take the system **BD+** of [12] which is obtained by adding Boolean complementation to **BD**, then the same trick, namely to replace the falsity condition for the conditional by the connexive one, will give us a functionally complete expansion of **BD**. Let us refer to the new expansion of **BD** as **dBD** (dialetheic **BD**). Semantically, **dBD**-valuation is obtained from **dLP**-valuation by dropping the exhaustivity condition $v^+(P) \cup v^-(P) = D^n$, and replacing the truth and falsity conditions for $\circ$ by the following conditions for $\neg$.

$$1 \in \overline{v}(\neg A) \text{ iff } 1 \notin \overline{v}(A),\ 0 \in \overline{v}(\neg A) \text{ iff } 0 \notin \overline{v}(A),$$

Then the truth tables for the propositional connectives become as follows.

A	$\sim A$	$\neg A$	$A \wedge B$	**t b n f**	$A \vee B$	**t b n f**	$A \to B$	**t b n f**
t	**f**	**f**	**t**	**t b n f**	**t**	**t t t t**	**t**	**t b n f**
b	**b**	**n**	**b**	**b b f f**	**b**	**t b t b**	**b**	**t b n f**
n	**n**	**b**	**n**	**n f n f**	**n**	**t t n n**	**n**	**b b b b**
f	**t**	**t**	**f**	**f f f f**	**f**	**t b n f**	**f**	**b b b b**

Note here that designated values are **t** and **b**. Proof theoretically, **dBD** is obtained from **dLP**′ by dropping (A10), and replacing (A¬3) by the following axiom:

$$\sim\neg A \leftrightarrow \neg\sim A \qquad (A\neg 3')$$

Based on these, the soundness and completeness results again carry over from **dLP**. Definitional completeness can be also proved in a similar manner. I only note here that $x \wedge \neg\sim x$, $\Delta(x \wedge \sim x)$, $\Delta(\neg x \wedge \neg\sim x)$, and $\neg x \wedge \sim x$ define $\delta_{\mathbf{t}}(x)$, $\delta_{\mathbf{b}}(x)$, $\delta_{\mathbf{n}}(x)$, and $\delta_{\mathbf{f}}(x)$ respectively, and $x \vee \neg x$, $(x \wedge \neg x) \to y$, $\neg((x \wedge \neg x) \to y)$ and $x \wedge \neg x$ define $\mathsf{C}_{\mathbf{t}}(x)$, $\mathsf{C}_{\mathbf{b}}(x)$, $\mathsf{C}_{\mathbf{n}}(x)$, and $\mathsf{C}_{\mathbf{f}}(x)$ respectively, where $\circ x$ and Δx are defined as $\neg(x \wedge \sim x) \wedge (x \vee \sim x)$ and $\circ(x \vee \mathsf{C}_{\mathbf{n}}(x))$ respectively.

4 Concluding Remarks

What I hope to have established in this paper are the following two points. First, I presented and explored an account of negation which realizes the understanding that Jaśkowski seems to have had in mind when he first formulated the problem of paraconsistency. One of the implications of buying the account of negation presented in this paper, is that the validity of the laws of double negation introduction and elimination becomes a necessary condition. As a future topic, I will investigate the more proof theoretic implications of the account given in this paper.

Second, I briefly motivated and introduced the notion of dialetheic logic, a special kind of paraconsistent logic that is expressive enough to represent dialetheia already in the level of propositional logic. Then I presented a new system of dialetheic logic that is definitionally complete, and thus Post complete, in its propositional fragment and expands the well-known system of paraconsistent logic **LP** by adding elements from different traditions in paraconsistency, namely Logics of Formal Inconsistency and connexive logic of Wansing. In brief, take Priest, and then first da Costize and second Wansingize to obtain a fully expressive paraconsistent logic! Moreover, I observed how to expand **BD** into a dialetheic and definitionally complete logic by adding the Boolean complementation and a connexive conditional.

As a next step, I will explore the naive set theory based on **dLP**. Here, I only note that non-triviality result is preserved, even if we take **dLP** as the underlying logic, as far as we formulate the axioms of naive set theory by making use of material conditional and biconditional based on paraconsistent negation. Moreover, we may formulate stronger extensionality principles with the help of detachable conditional, and we may still prove the non-triviality result.

Note finally, that if we take **dBD** in developing naive set theory, then there is a problem to be faced. More specifically, if one sticks to the conditional and biconditional based on paraconsistent negation, then we lose the intuitive reading of biconditional. Indeed, in **LP**-based setting, the biconditional $A \equiv B$ is true iff A and B are both true or both false, but this will no longer be the case once we move to **BD**-based theories. Note also that one will be in trouble if one tries to redefine the biconditional as $(A \wedge B) \vee ({\sim}A \wedge {\sim}B)$ to keep the intuitive reading of the biconditional. This is because $(A \wedge \neg A) \vee ({\sim}A \wedge {\sim}\neg A)$ leads us to triviality in view of (A¬3'). What to say in this case remains to be seen.

Appendix

Details of Remark 10 Consider the algebra $\langle\{\mathbf{t}, \mathbf{b}, \mathbf{f}\}, \{\bot, \sim, \wedge, \vee, \rightarrow\}\rangle$ where the operations are defined as follows:

A	$\bot$	$\sim A$	$A \wedge B$	**t**	**b**	**f**	$A \vee B$	**t**	**b**	**f**	$A \rightarrow B$	**t**	**b**	**f**
t	**f**	**f**	**t**	**t**	**b**	**f**	**t**	**t**	**t**	**t**	**t**	**t**	**b**	**f**
b	**f**	**b**	**b**	**b**	**b**	**f**	**b**	**t**	**b**	**b**	**b**	**t**	**b**	**f**
f	**f**	**t**	**f**	**f**	**f**	**f**	**f**	**t**	**b**	**f**	**f**	**b**	**b**	**b**

The aim here is to show that $\circ$ is not definable in this algebra. To this end, we prove the following lemma.

Lemma 4 *Let $\varphi(p)$ be any formula in the language $\mathcal{L}_\bot$ whose only propositional variable is p. Then, there are seven cases for the value of $\varphi(p)$ depending on the* *value assigned to p, namely:*

p	(1)	(2)	(3)	(4)	(5)	(6)	(7)
t	**t**	**t**	**b**	**b**	**b**	**f**	**f**
b	**t**	**b**	**t**	**b**	**f**	**b**	**f**

Proof We proceed by induction on the complexity of $\varphi(p)$. For the base case, if $\varphi(p)$ is p or $\perp$, then it satisfies the condition (2) or (7) respectively. For the induction step, we cover only three of the four cases, as the others are similar.

Case 1: let $\varphi(p)$ be of the form $\sim\psi(p)$. Then, by induction hypothesis, $\psi(p)$ satisfies one of the seven cases. And with the truth table for $\sim$ in mind, $\varphi(p)$ satisfies the condition $(8-i)$ when $\psi(p)$ satisfies (i) $(i \in \{1, 2, \ldots, 7\})$ respectively.

Case 2: let $\varphi(p)$ be of the form $\psi(p) \wedge \xi(p)$. Then, by induction hypothesis, $\psi(p)$ and $\xi(p)$ both satisfy one of the eight conditions. And with the truth table for $\wedge$ in mind, $\varphi(p)$ behaves as follows:

$\psi(p)\wedge\xi(p)$	(1)	(2)	(3)	(4)	(5)	(6)	(7)
(1)	(1)	(2)	(3)	(4)	(5)	(6)	(7)
(2)	(2)	(2)	(4)	(4)	(5)	(6)	(7)
(3)	(3)	(4)	(3)	(4)	(5)	(6)	(7)
(4)	(4)	(4)	(4)	(4)	(5)	(6)	(7)
(5)	(5)	(5)	(5)	(5)	(5)	(7)	(7)
(6)	(6)	(6)	(6)	(6)	(7)	(6)	(7)
(7)	(7)	(7)	(7)	(7)	(7)	(7)	(7)

The case for disjunction is similar to the case for conjunction.
Case 3: let $\varphi(p)$ be of the form $\psi(p) \rightarrow \xi(p)$. Then, by induction hypothesis, $\psi(p)$ and $\xi(p)$ both satisfy one of the eight conditions. And with the truth table for $\wedge$ in mind, $\varphi(p)$ behaves as follows:

$\psi(p) \rightarrow \xi(p)$	(1)	(2)	(3)	(4)	(5)	(6)	(7)
(1)	(1)	(2)	(3)	(4)	(5)	(6)	(7)
(2)	(1)	(2)	(3)	(4)	(5)	(6)	(7)
(3)	(1)	(2)	(3)	(4)	(5)	(6)	(7)
(4)	(1)	(2)	(3)	(4)	(5)	(6)	(7)
(5)	(3)	(4)	(3)	(4)	(4)	(5)	(5)
(6)	(2)	(2)	(4)	(4)	(6)	(4)	(6)
(7)	(4)	(4)	(4)	(4)	(4)	(4)	(4)

This completes the proof. □

This implies that $\circ$ is not definable. Indeed, if $\circ$ is definable, then we will have the case in which $\circ$**t** = **t** and $\circ$**b** = **f**, but this is not the case in view of the above lemma.

Details of Remark 20 In view of Theorem 2, and that $\to$ is essentially binary, it suffices to show that all unary functions are definable in the algebra $\langle\{\mathbf{t}, \mathbf{b}, \mathbf{f}\}, \{\sim, \circ, \to\}\rangle$. In other words, we need to show $27(= 3^3)$ functions are definable. This can be done as follows.

$\circ\circ x$	$\sim(x\to\sim\circ\circ x)$	$\circ(x\to\sim\circ\circ x)$	$\sim(\circ x\to\sim\circ x)$	$x\to x$	x	$\circ x$	$x\to\circ x$	$\circ(x\to x)$
t	**t**	**t**	**t**	**t**	**t**	**t**	**t**	**t**
t	**t**	**t**	**b**	**b**	**b**	**f**	**f**	**f**
t	**b**	**f**	**t**	**b**	**f**	**t**	**b**	**f**

$\sim\circ(x\to x)\to\circ\circ x$	$\sim(\sim\circ x\to\circ x)$	$\sim\circ(x\to x)\to\sim\circ x$	$\sim(\sim x\to x)$	$\sim\circ\circ x\to x$
b	**b**	**b**	**b**	**b**
t	**t**	**t**	**b**	**b**
t	**b**	**f**	**t**	**b**

$\sim x\to x$	$\sim\circ(x\to x)\to\circ x$	$\sim\circ x\to\circ x$	$\sim\circ(x\to x)\to\sim\circ\circ x$
b	**b**	**b**	**b**
b	**f**	**f**	**f**
f	**t**	**b**	**f**

$\sim\circ(x\to x)$	$\sim(x\to\circ x)$	$\sim\circ x$	$\sim x$	$\sim(x\to x)$	$\circ x\to\sim\circ x$	$\sim\circ(x\to\sim\circ\circ x)$	$x\to\sim\circ\circ x$	$\sim\circ\circ x$
f	**f**	**f**	**f**	**f**	**f**	**f**	**f**	**f**
t	**t**	**t**	**b**	**b**	**b**	**f**	**f**	**f**
t	**b**	**f**	**t**	**b**	**f**	**t**	**b**	**f**

This completes the proof. □

References

1. Arieli, O., & Avron, A. (1998). The value of the four values. *Artificial Intelligence*, *102*, 97–141.
2. Avron, A. (1999). On the expressive power of three-valued and four-valued languages. *Journal of Logic and Computation*, *9*, 977–994.
3. Batens, D. (1980). Paraconsistent extensional propositional logics. *Logique et Analyse*, *90–91*, 195–234.
4. Batens, D., & De Clercq, K. (2004). A rich paraconsistent extension of full positive logic. *Logique et Analyse*, *185–188*, 227–257.
5. Beall, J. C. (2009). *Spandrels of truth*. Oxford: Oxford University Press.
6. Cantwell, J. (2008). The logic of conditional negation. *Notre Dame Journal of Formal Logic*, *49*, 245–260.
7. Carnielli, W., Coniglio, M., & Marcos, J. (2007). Logics of formal inconsistency. In D. Gabbay & F. Guenthner (Eds.), *Handbook of philosphical logic* (Vol. 14, pp. 1–93). Dordrecht: Springer.
8. Carnielli, W. & Marcos, J. (2002). A Taxonomy of C-systems. In W. A. Carnielli, M. E. Coniglio, & I. M. L. d'Ottaviano, (Eds.), *Paraconsistency: the logical way to the inconsistent*, (pp. 1–94). New York: Marcel Dekker.
9. Carnielli, W., Marcos, J., & de Amo, S. (2000). Formal inconsistency and evolutionary databases. *Logic and Logical Philosophy*, *8*, 115–152.

10. Ciuciura, J. (2008). Frontiers of the discursive logic. *Bulletin of the Section of Logic*, *37*(2), 81–92.
11. da Costa, & Newton, C. A. (1974). On the theory of inconsistent formal systems. *Notre Dame Journal of Formal Logic*, *15*, 497–510.
12. de Michael, & Omori, H. (2015). Classical negation and expansions of BD. *Studia Logica*, *103*(4), 825–851.
13. d'Ottaviano, I. M. L. (1985).The completeness and compactness of a three-valued first-order logic. *Revista Colombiana de Matemáticas*, *19*:77–94,
14. d'Ottaviano, I. M. L., Newton C. A. & da Costa. (1970). Sur un problème de jaśkowski. *Comptes Rendus de l'Academie de Sciences de Paris (A-B)*, *270*:1349–1353
15. Goodship, L. (1996). On dialethism. *Australasian Journal of Philosophy*, *74*(1), 153–161.
16. Horn, L. R., & Wansing, H. (2015). Negation. In E. N. Zalta (Ed.), *The stanford encyclopedia of philosophy*. http://plato.stanford.edu/entries/negation/ (Spring 2015 edition).
17. Jaśkowski, S. (1999). A propositional calculus for inconsistent deductive systems. *Logic and Logical Philosophy*, *7*:35–56 (A new translation based on [?])
18. Jaśkowski, S. (1999). On the discussive conjunction in the propositional calculus for inconsistent deductive systems. *Logic and Logical Philosophy*, *7*, 57–59.
19. Kamide, N., & Wansing, H. (2012). Proof theory of nelson's paraconsistent logic: a uniform perspective. *Theoretical Computer Science*, *415*, 1–38.
20. Lenzen, W. (1996). Necessary conditions for negation operators. In Wansing, H. (Ed.), *Negation: a notion in focus*, Perspective in Analytical Philosophy (pp. 37–58). Walter de Gruyter,
21. Marcos, J. (2005). On negation: pure local rules. *Journal of Applied Logic*, *3*, 185–219.
22. McCall, S. (2012). A history of connexivity. In D. M. Gabbay, F. Pelletier, & J. Woods (Eds.), *Logic: a history of its central concepts*, Handbook of the history of logic, (Vol. 11, pp. 415–449). Elsevier
23. Muskens, R. (1999). On partial and paraconsistent logics. *Notre Dame Journal of Formal Logic*, *40*(3), 352–374.
24. Olkhovikov, G. (2001). On a new three-valued paraconsistent logic (in Russian). In *Logic of Law and Tolerance* (pp. 96–113). Yekaterinburg: Ural State University Press
25. Omori, Hitoshi. (2015). Remarks on naive set theory based on LP. *The Review of Symbolic Logic*, *8*(2), 279–295.
26. Omori, H., & Sano, K. (2014). da Costa meets Belnap and Nelson. In R. Ciuni, H. Wansing, & C. Willkommen (Eds.), *Recent Trends in Philosophical Logic* (pp. 145–166). Berlin: Springer.
27. Omori, H., & Sano, K. (2015). Generalizing Functional Completeness in Belnap-Dunn Logic. *Studia Logica*, *103*(5), 883–917.
28. Omori, H., & Waragai, T. (2011). Some Observations on the Systems **LFI1** and **LFI1***. *Proceedings of Twenty-Second International Workshop on Database and Expert Systems Applications (DEXA2011)* (pp. 320–324)
29. Omori, H. & Waragai, T. (2014). On the propagation of consistency in some systems of paraconsistent logic. In E. Weber, D. Wouters, & J. Meheus, (Eds.), *Logic, reasoning and rationality*, (pp. 153–178). Heidelberg: Springer
30. Priest, G. (2002). *Beyond the limits of thought* (2nd ed.). Oxford: Oxford University Press.
31. Priest, G. (2006). *Doubt truth to be a liar*. Oxford: Oxford University Press.
32. Priest, G. (2006). *Contradiction* (2nd ed.). Oxford: Oxford University Press.
33. Priest, G. (2014). *One*. Oxford: Oxford University Press.
34. Priest, G., & Routley, R. (1989). Systems of paraconsistent logic. In G. Priest, R. Routley, & J. Norman (Eds.), *Paraconsistent logic: essays on the inconsistent* (pp. 151–186). Munich: Philosophia
35. Pynko, A. P. (1999). Functional completeness and axiomatizability within Belnap's four-valued logic and its expansions. *Journal of Applied Non-Classical Logics*, *9*(1), 61–105.
36. Restall, G. (1992). A note on naive set theory in LP. *Notre Dame Journal of Formal Logic*, *33*, 422–432.
37. Ruet, P. (1996). Complete set of connectives and complete sequent calculus for Belnap's logic. Technical report, Ecole Normale Superieure, *Logic Colloquium* 96, Document LIENS-96-28.

38. Słupecki, J. (1972). A criterion of fullness of many-valued systems of propositional logic. *Studia Logica*, *30*, 153–157.
39. Tokarz, M. (1973). Connections between some notions of completeness of structural propositional calculi. *Studia Logica*, *32*(1), 77–89.
40. Wansing, H. (2001). Negation. In L. Goble (Ed.), *The blackwell guide to philosophical logic* (pp. 415–436). Cambridge: Basil Blackwell Publishers.
41. Wansing, H. (2005). Connexive modal logic. In R. Schmidt, I. Pratt-Hartmann, M. Reynolds & H. Wansing, (Eds.), *Advances in modal logic* (Vol. 5, pp. 367–383). London: King's College Publications
42. Wansing, H. (2014). Connexive logic. In E. N. Zalta (Ed.), *The Stanford Encyclopedia of Philosophy*. http://plato.stanford.edu/archives/fall2014/entries/logic-connexive/ (Fall 2014 edition)
43. Weber, Z. (2010). Transfinite numbers in paraconsistent set theory. *The Review of Symbolic Logic*, *3*(1), 71–92.
44. Weber, Z. (2012). Transfinite cardinals in paraconsistent set theory. *The Review of Symbolic Logic*, *5*(2), 269–293.

Paradoxes of Expression

Martin Pleitz

Abstract In this note, I show how to construct Liar-like and Curry-like paradoxes in a framework Graham Priest has been considering recently, in which he tries to solve the paradoxes by giving up the rule of modus ponens (detachement) instead of the rules of ex falso and contraction. The Curry-like paradox presents a serious challenge to the detachment-free framework because it threatens to trivialize the system, just as Curry's original paradox does for the more standard paraconsistent approach to the paradoxes.

Graham Priest in some recent talks and unpublished work investigates the possibility, also discussed favorably but in less detail in Goodship [4] and defended in Beall [2], of solving the semantic and set theoretic paradoxes by restricting the truth schema, the naive set abstraction schema, and so on by formulating them with a biconditional that does not *detach*, i.e., that does not satisfy modus ponens [12]. But a detachable truth schema is needed for the usual account of *blind endorsement*, i.e., of the ascription of truth to sentences that are identified in a way that gives no clue about their content (e.g., when someone holds that everything the Bible says is true). To solve this problem, Priest deliberates whether to add a further, detachable conditional, propositional quantifiers, and an "expression predicate" (Priest [12], Sect. 5.2). In this note, I will show that given a natural principle about expression, both Liar-like and Curry-like paradoxes can then be constructed without any appeal to the truth schema or to its relatives. I will discuss briefly what these paradoxes of expression mean for such detachment-free approaches to paradox.[1]

Given Priest's usual paraconsistent approach to paradox that is based on the logic LP (Priest [11], 53ff.), it is natural for him to work with a non-detachable biconditional, as LP does not in general sanction the move from '$p \supset q$' and 'p' to 'q' if

[1]I would like to thank an anonymous referee, Johannes Korbmacher, Tobias Martin, Graham Priest, Stewart Shapiro, and Niko Strobach for helpful discussions and comments.

M. Pleitz (✉)
Philosophisches Seminar, Westfälische Wilhelms-Universität,
Domplatz 6, D-48143 Müenster, Germany
e-mail: martinpleitz@web.de

H. Andreas and P. Verdée (eds.), *Logical Studies of Paraconsistent Reasoning in Science and Mathematics*, Trends in Logic 45,
DOI 10.1007/978-3-319-40220-8_9

'$\supset$' is defined in the usual way from negation and disjunction or conjunction. And he explores some metaphysical options that open up when the theory of identity is based on a non-detachable conditional in his book on unity (Priest [13], 16ff.). In his usual approach to the paradoxes, however, the Liar and its ilk are not solved by restricting the truth schema, but by dropping the ex falso rule, so that it becomes acceptable that there are (some) *dialetheias*, i.e., true contradictions. Later on, Priest and other dialetheists normally add a detachable conditional (Priest [11], 82ff.), probably because modus ponens is held in high regard even among the most radically revisionist logicians. This additional conditional must not obey the rule of contraction, though, because otherwise Curry's paradox would trivialize the system (Meyer/Routley/Dunn [6]; Beall [1], Sect. 3.1).

As Laura Goodship points out, the alternative strategy of solving the paradoxes not by giving up ex falso to deal with paradoxes like the Liar and the rule of contraction to deal with Curry's paradox, but by giving up modus ponens for the conditional that figures in the truth schema etc., has the advantage of solving the Liar and Curry's paradox in a *uniform* way (Goodship [4], 157ff.; Priest [12], Sect. 4.1). A uniform solution is desirable if these paradoxes are of the same kind (Priest [10], 166f.) – which arguably they are (Grattan-Guinness [5], 826f.; Pleitz [7]).

So, despite the price of giving up modus ponens, a detachment-free approach to the paradoxes is worth investigating. According to this approach, the biconditional '$\equiv$' in the *truth schema*[2]

$$\text{TRUE}(\underline{\alpha}) \equiv \alpha$$

is understood as not obeying the rule of modus ponens.[3] This restriction of the schema blocks the paradoxical arguments of both the Liar and Curry, all of which employ modus ponens at one point or another.

But Priest thinks that there is a problem for this proposal according to which "truth does not detach" that is connected to a role the truth predicate is often thought to play – the role of making expressible *blind endorsement*, as for instance in the statement that everything that the Bible says is true (Priest [12], Sect. 3.2). Priest wants to accommodate arguments like the following, which move from some such blind endorsement to the fact that a certain state of affairs obtains:

P1 Everything that the Bible says is true.
P2 The Bible says: 'For every purpose, there is a season.'
C So, for every purpose, there is a season.

Standardly, this is understood as elliptical for an argument that makes use of an instance of the truth schema:

---

[2] I use underlining to indicate a name-forming operator.

[3] Although Priest thinks of it as the biconditional of LP, he calls it the "material biconditional". I find this terminology misleading, because when the conditional of LP is contrasted with some conditional that *does* detach, the latter need not at all be intensional and might even be the material biconditional of classical logic.

P1 Everything that the Bible says is true.
P2 The Bible says: 'For every purpose, there is a season.'
C1 'For every purpose, there is a season.' is true.
P3 'For every purpose, there is a season.' is true iff …
… for every purpose, there is a season.
C2 For every purpose, there is a season.

This argument, however, makes use of modus ponens (in the move from C1 and P3 to C2, and – given the usual formalization of restricted quantification – already in the move from P1 and P2 to C1). So, if the 'if and only if' of the truth schema (here in P3) does not detach, the argument is not valid.

Priest thinks about solving this problem by adding to the system he discusses (where the truth schema is formulated with a biconditional that does not detach) propositional quantification and a further, detachable conditional '$\rightarrow$'. This would amount to "cutting out the middle man [i.e., the truth predicate] entirely" (Priest [12], Sect. 5.2). Using '$B : p$' as a sentential operator with the intended meaning 'The Bible says that p' and the propositional constant 'p_0' for the proposition that for every purpose, there is a season, the original argument can now be formalized in the following way:

1	$\forall p(B : p \rightarrow p)$	P1; blind endorsement
2	$B : p_0 \rightarrow p_0$	from 1 by $\forall$-Elim
3	$B : p_0$	P2; what the Bible says
4	p_0	from 2 and 3 by MP

As the conditional '$\rightarrow$' detaches, this argument is valid. The truth schema was not employed at all, so this explication is compatible with the truth schema being formulated with a non-detachable biconditional. The problem that detachment-free approaches to paradox had with accounting for blind endorsement thus seems to be solved.

* * *

But things are more complicated. It will turn out that if in addition to propositional quantification and the further, detachable conditional '$\rightarrow$' and the corresponding detachable biconditional '$\leftrightarrow$', there is an expression predicate-operator that satisfies a very natural principle, then both a counterpart of the Liar paradox and a counterpart of Curry's paradox can be recovered in the system discussed by Priest.

The expression predicate-operator is '$E(x : p)$', with the intended meaning 'x expresses that p'. Here, 'x' can be replaced by a term (a constant or a variable) that refers to a sentence (or another kind of object) and 'p' can be replaced by a particular proposition (i.e., a sentence of the language) or a propositional constant or variable. So '$E(\ldots : \ldots)$' is a predicate with regard to the left hand side of the colon and a sentential operator with regard to the right hand side of the colon (which is why I call it a *predicate-operator*).

The expression predicate-operator can be used to define other notions in the vicinity[4]:

(D1) $\exists p\ E(x : p)$	the sentence x is *meaningful*
(D2) $\exists x\ E(x : p)$	the proposition that p is *expressible*
(D3) $\forall p(E(x : p) \leftrightarrow E(y : p))$	the sentences x and y are *synonymous* (if meaningful)

For the following recovery of the Liar and Curry's paradox, let us think of the expression predicate-operator as governed by a single axiom:

$$\text{(E)} \quad \forall x \forall p \forall q(E(x : p) \wedge E(x : q) \rightarrow (p \leftrightarrow q))$$

Note that, if definition (D3) does indeed capture the notion of synonymity adequately, then (E) says no more than that every (meaningful) sentence is synonymous to itself.[5] As modus ponens holds for the conditional '$\rightarrow$', the axiom (E) sanctions the inference from the three premises '$E(x : p)$', '$E(x : q)$', and 'p' to the conclusion 'q'. It's worth noting that the following conundrum could also be formulated on the basis of the requirement that the expression predicate-operator, instead of the axiom (E), satisfies this inferential rule.[6] So the question will really be: Is *expression* detachable?

Now for the paradoxes. By making use of propositional quantification and the expression predicate-operator, we can formulate that what a certain sentence says is not the case, which comes close to saying that it is false, and we can formulate that everything a certain sentence says entails a certain proposition, which comes close to saying that the truth of that sentence entails that proposition. Given some device that can make a sentence self-referential, there would thus be sentences much like a Liar sentence and a Curry sentence.[7] Their existence could for instance be stipulated by the following identity statements:

$$(\text{L}_=) \quad l = \underline{\forall p(E(\underline{l} : p) \rightarrow \neg p)}$$

$$(\text{C}_=) \quad c = \underline{\forall p(E(\underline{c} : p) \rightarrow (p \rightarrow q_0))}$$

In the presence of the expression predicate-operator, however, we do not need the usual devices of the identity symbol and a name-forming operator to claim that there

---

[4] I call '$E(x : p)$' not an 'expressibility' but an '*expression*' predicate-operator because, given the intended meaning of '$E(x : p)$', the notion of *expressibility* is captured much better by the defined operator '$\exists x\ E(x : p)$'; cf. (D2).

[5] In my discussion of the paradoxes of expression, I will consider and reject an objection to (E) that is based on the observation that some sentences are ambiguous.

[6] However, the conditional that obeys modus ponens would still be needed for the formulation of the Liar and Curry sentences.

[7] This way of formalizing a Liar sentence via propositional quantification and negation is not original. Cf., e.g., Prior [14]. But to my knowledge, it has not yet been transferred to Curry's paradox.

is a certain self-referential sentence. We need only say of a certain sentence that it expresses that *it* has a certain property. More specifically, to stipulate the existence of a Liar sentence, we need only claim that there is a sentence l that expresses that for every proposition p expressed by l, it is not the case that p. And to stipulate the existence of a Curry sentence, we need only claim that there is a sentence c which expresses that for every proposition p expressed by c, if p then q_0 (where 'q_0' is an arbitrary propositional constant). Formally:

$$\text{(L)}\quad E(l : \forall p(E(l : p) \rightarrow \neg p))$$

$$\text{(C)}\quad E(c : \forall p(E(c : p) \rightarrow (p \rightarrow q_0)))$$

The following derivation shows that l behaves like a Liar sentence insofar it allows to infer a contradiction.

(E)		$\forall x \forall p \forall q(E(x : p) \wedge E(x : q) \rightarrow (p \leftrightarrow q))$	a property of expression
(L)		$E(l : \forall p(E(l : p) \rightarrow \neg p))$	the existence of a Liar sentence
1	*	$E(l : p_0)$	assumption (for CP)
2	*	$E(l : p_0) \wedge E(l : \forall p(E(l : p) \rightarrow \neg p)) \rightarrow \ldots$	(E), ∀-Elim
		$\ldots (p_0 \leftrightarrow \forall p(E(l : p) \rightarrow \neg p))$	
3	*	$p_0 \leftrightarrow \forall p(E(l : p) \rightarrow \neg p)$	(L), 1, 2, MP
4	**	p_0	assumption (for RA)
5	**	$\forall p(E(l : p) \rightarrow \neg p)$	4, 3, MP
6	**	$E(l : p_0) \rightarrow \neg p_0$	5, ∀-Elim
7	**	$\neg p_0$	1, 6, MP
8	*	$\neg p_0$	4–7, RA (¬-Intro)
9		$E(l : p_0) \rightarrow \neg p_0$	1–8, CP
10		$\forall p(E(l : p) \rightarrow \neg p)$	9, ∀-Intro
11		$\exists r E(l : r)$	(L), ∃-Intro
12		$E(l : r_0)$	11, ∃-Elim
13		$E(l : r_0) \wedge E(l : \forall p(E(l : p) \rightarrow \neg p)) \rightarrow \ldots$	(E), ∀-Elim
		$\ldots (r_0 \leftrightarrow \forall p(E(l : p) \rightarrow \neg p))$	
14		$r_0 \leftrightarrow \forall p(E(l : p) \rightarrow \neg p)$	(L), 12, 13, MP
15		r_0	10, 14, MP
16		$E(l : r_0) \rightarrow \neg r_0$	10, ∀-Elim
17		$\neg r_0$	12, 16, MP
18		$r_0 \wedge \neg r_0$	14, 17, ∧-Intro

Because of line 12 and (E), r_0 can be seen as the unique proposition expressed by the Liar sentence l (modulo the detachable biconditional '$\leftrightarrow$').[8] Hence the result in line 18 is very close to the conclusion of the usual argument of the Liar paradox: What the Liar sentence expresses is and is not the case.

[8]Note that, as 'l' is an individual constant that can only stand in the position of a singular term, '$l \wedge \neg l$' is ill-formed and cannot be used to express the result of the Liar reasoning. 'r_0', in contrast, is a *propositional* constant so that '$r_0 \wedge \neg r_0$' is well-formed.

The following derivation shows that c behaves like a Curry sentence insofar it allows to infer the arbitrary proposition q_0.

(E)		$\forall x \forall p \forall q(E(x:p) \wedge E(x:q) \rightarrow (p \leftrightarrow q))$	a property of expression
(C)		$E(c: \forall p(E(c:p) \rightarrow (p \rightarrow q_0)))$	the existence of a Curry sentence
1	*	$E(c:p_0)$	assumption (for CP)
2	*	$E(c:p_0) \wedge E(c: \forall p(E(c:p) \rightarrow (p \rightarrow q_0))) \rightarrow \ldots$ $\ldots (p_0 \leftrightarrow \forall p(E(c:p) \rightarrow (p \rightarrow q_0)))$	(E), ∀-Elim
3	*	$p_0 \leftrightarrow \forall p(E(c:p) \rightarrow (p \rightarrow q_0))$	(C), 1, 2, MP
4	**	p_0	assumption (for CP)
5	**	$\forall p(E(c:p) \rightarrow (p \rightarrow q_0))$	4, 3, MP
6	**	$E(c:p_0) \rightarrow (p_0 \rightarrow q_0)$	5, ∀-Elim
7	**	$p_0 \rightarrow q_0$	1, 6, MP
8	**	q_0	4, 7, MP
9	*	$p_0 \rightarrow q_0$	4–8, CP
10		$E(c:p_0) \rightarrow (p_0 \rightarrow q_0)$	1–9, CP
11		$\forall p(E(c:p) \rightarrow (p \rightarrow q_0))$	10, ∀-Intro
12		$\exists r E(c:r)$	(C), ∃-Intro
13		$E(c:r_0)$	12, ∃-Elim
14		$E(c:r_0) \wedge E(c: \forall p(E(l:p) \rightarrow (p \rightarrow q_0))) \rightarrow \ldots$ $\ldots (r_0 \leftrightarrow \forall p(E(c:p) \rightarrow (p \rightarrow q_0)))$	(E), ∀-Elim
15		$r_0 \leftrightarrow \forall p(E(c:p) \rightarrow (p \rightarrow q_0))$	(C), 13, 14, MP
16		r_0	11, 15, MP
17		$E(c:r_0) \rightarrow (r_0 \rightarrow q_0)$	11, ∀-Elim
18		$r_0 \rightarrow q_0$	13, 17, MP
19		q_0	16, 18, MP

Note that it is essential to these derivations that the conditional in the Liar sentence that is stipulated to exist by (L) and both conditionals in the Curry sentence that is stipulated to exist by (C) are detachable.[9] But it will not help the paradox solver to question whether the conditionals in (L) and (C) should rather be the non-detachable ones used for the truth schema. Because once the possibility of self-reference is granted,[10] there will be all kinds of self-referential sentences; and although many of them – including detachment-free variants of l and c – might be harmless, there will *also* be l and c, which lead to trouble.

The two derivations show that the combined resources of propositional quantification and an expression predicate-operator that conforms to the principle (E) suffice to produce Liar-like and Curry-like paradoxes. This is not too big a surprise, as these resources allow to define a (detachable) truth predicate that applies to sentences by introducing 'TRUE(x)' as an abbreviation for '$\exists p\ E(x:p) \wedge \forall p(E(x:p) \rightarrow p)$'

[9]This is witnessed by line 7 of the Liar derivation and line 7 and 8 of the Curry derivation.

[10]One could (and I personally would) draw an entirely different moral from the paradoxes of expression and all other paradoxes that are based on self-referential expressions, namely that the required kind of self-referential expressions which attribute semantic properties to themselves do not exist, after all; cf. Pleitz ([8]). But in the present context of a discussion within the horizon of those paradox-solvers who have taken the path of logical revision, the existence of the problematical self-referential expressions is of course a given.

(or for '$\exists p(E(x : p) \wedge p)$', which in the presence of (E) is equivalent).[11] Given this formalization of truth, (L) says that the Liar sentence l says of itself that it is not true if meaningful. And as (L) further entails that l is meaningful (cf. line 11 in the first derivation), this come close enough to l saying of itself that it is not true. In a similar way, the second derivation shows that c comes close to saying of itself that if it is true, then q_0. But this illustration in terms of truth is only by way of an intuitive understanding. For the assessment of these paradoxes in the context of the detachment-free approach to paradox discussed by Priest it is important to keep in mind that it is not the notion of truth, but the notion of expression that leads to trouble here.

* * *

For an approach to the paradoxes that is based on the idea that truth does not detach, these results are problematic. More specifically, as soon as we follow Priest in formally explicating blind endorsement by adding propositional quantifiers and a further, detachable conditional to the machinery with the non-detachable truth schema, new Liar-like and Curry-like paradoxes recur as soon as a predicate-operator of expression is added. The dialetheist can of course accept the Liar-like paradoxes (and even see them as adding grist to his or her mill) because they produce only a few further true contradictions. But the dialetheist cannot stomach the Curry-like paradoxes, because they will trivialize the system. The paradoxes of expression could of course be solved by going paraconsistent and contraction-free, but then the truth schema might as well be formulated with a contraction-free conditional that *does* obey modus ponens, and the dialetheist would be back with Priest's non-uniform solution of *In Contradiction*.

So, do we really need to add an expression predicate-operator? Does it really have to conform to principle (E)? I think that both questions should be answered in the affirmative.

The introduction of a predicate-operator of expression can arguably be motivated already from the Bible example; in the following way. While the phrase 'everything in the Bible' is most naturally understood to range over *sentences*, the machinery of propositional quantification calls for expressions that allow to talk about *propositions* to formalize the phrase 'is true', e.g. like this: Let '$B(x)$' be a predicate of sentences with the intended meaning 'x is a sentence of the Bible'. Then the sentence 'Everything in the Bible is true' gets formalized as '$\forall x(B(x) \rightarrow (\exists p\, E(x : p) \wedge \forall p(E(x : p) \rightarrow p)))$'.

More generally, and more to the point, it can be argued that it is the specific characteristics of the system discussed by Priest *themselves* which call for the addition of the expression predicate-operator. As the main purpose of this system is to

[11] Priest makes a brief remark about an expression '$E(x, p)$', which he calls a "binary predicate" (towards the end of Sect. 5.2 of Priest [12]), that is similar in intended meaning to our expression predicate-operator '$E(x : p)$'. He proposes to define a truth predicate from it as '$\exists p(p \wedge E(x, p))$' or as '$\forall p\ (E(x, p) \supset p)$'. I note that these are not equivalent, because in contrast to the first, the second of these two open formulas will be vacuously satisfied by anything that does not express something.

solve paradoxes like the Liar, it needs to have resources to talk about sentences (so that there are Liar sentences to start with). Priest's emendation of the system with propositional quantification allows to quantify into sentence position and – via propositional variables and constants – to talk about what sentences express, i.e., propositions. Hence it would be quite unnatural if the language could not express whether a certain sentence expresses a certain proposition.[12]

Even *more* generally, it would hardly be in the spirit of the dialetheist approach to the paradoxes, which tries to regain the semantic closure surrendered by classical and many other approaches to paradox, if it needed to ban a resource that allows to express expression.

With regard to the crucial principle (E), let me repeat first that because it is formulated with a detachable conditional, it justifies the inference from 'x expresses that p and x expresses that q' and 'p' to 'q', and that this rule of inference alone would suffice for the above paradoxical reasoning. In contrast, changing principle (E) by replacing the detachable biconditional '$\leftrightarrow$' with the non-detachable biconditional '$\equiv$' would remove the justification from this form of argument. So the question is really whether there is anything objectionable about these inferences. To me they seem even more intuitive than the argument from the truth of everything the Bible says to the conclusion that for every purpose, there is a season.

One possible objection focuses on the issue of *ambiguity*. Someone might think that the sentence 'Sarah goes to the bank' expresses both that Sarah goes to the side of the river and that Sarah goes to the money institute, and even though they know that Sarah goes to the river-side resist the inference to the conclusion that she goes to a money institute. But this example would be relevant to the question at hand only if 'x expresses p' were construed as expressing a *relation*, and quantification into the second place of this dyadic expression would not be *propositional quantification*, but first-order quantification over propositions construed as *objects* (maybe in a language with many-sorted quantification). For note that propositional quantification – that is, quantification *into sentence position* – does not enable us to speak about propositions

[12]With a view of recovering the paradoxes in the system discussed by Priest, we could put the predicate-operator '$E(x : p)$' to the side, and work in its stead with a unary expression-operator
(Footnote 12 continued)
for each one of the problematic sentences, e.g., '$L(p)$' with the intended meaning that the specific sentence l expresses that p, and '$C_n(p)$' with the intended meaning that the specific sentence c_n expresses that p. In parallel to the stipulations (L) and (C) above, we could lay down that $L(\forall p(L(p) \rightarrow \neg p)$, that $C_n(\forall p(C_n(p) \rightarrow (p \rightarrow q_0)))$, and so on, and would thus guarantee that l would be a Liar sentence, c_n would be a Curry sentence, and so on. Now, given principles much like the respective instances of (E) for each one of the sentences associated with these unary operators – e.g., $\forall p \forall q(L(p) \wedge L(q) \rightarrow (p \leftrightarrow q))$ for the operator '$L(...)$' that concerns sentence l, – counterparts of the above derivations would be valid. I would like to thank an anonymous referee for alerting me to this possibility. But I am not convinced that much would be gained for our discussion of Priest's system by the ensuing proliferation of unary operators, each governed by its own specific principle. In view of the dialectic of the debate and the aim of testing the system discussed by Priest, it is important to introduce resources that capture a detaching notion of expression in a way that is motivated independently of the paradoxes. And, in contrast to the principles governing the specific operators needed for the paradoxical derivations, there are general considerations that provide such independent motivation in the case of principle (E).

that are more fine-grained than the sentences that express them! In the example concerning the word 'bank', we smuggled in resources to distinguish two meanings of one sentence (the phrases 'river-side' and 'money institute'), but this move cannot be formalized in terms of those symbols that allow to talk about propositions in a language with propositional quantification. If E('Sarah goes to the bank': p_0), then the proposition p_0 arguably is the *ambiguous* sentence meaning that Sarah goes to the river-side / the money institute.

Note also that even if ambiguity *did* constitute a good reason to restrict principle (E), this need not change anything for Liar sentences and Curry sentences. For why should *they* be ambiguous?[13] To construe Liar sentences and Curry sentences as ambiguous would move us into the vicinity of Stephen Read's Bradwardinian solution to the paradoxes [15], which in present terms amounts to accepting the above definition of truth, but rejecting principle (E).

In sum, Priest's emendation of the detachment-free proposal to solve the paradoxes by the addition of propositional quantification and a detachable conditional to formalize blind endorsement does not fare well. As there are strong reasons to also add an expression predicate-operator that is in general governed by principle (E), the above variant of Curry's paradox threatens to trivialize the system. At this point, a friend of the detachment-free proposal can still go two ways: reject contraction for the additional, detachable conditional, or argue that a Curry sentence says several different things. But either way will take him or her away from the detachment-free proposal. The first way leads back to Priest's proposal in *In Contradiction*, and the second way leads to Read's Bradwardinian solution.

References

1. Beall, J. (2009). Curry's paradox. In E. N. Zalta (Ed.), *The Stanford Encyclopedia of Philosophy* (Spring 2009 Edition). http://plato.stanford.edu/archives/spr2009/entries/curry-paradox/.
2. Beall, J. (2013). Free of detachment: logic, rationality, and gluts. *Noûs*, *49*(1), 1–14.
3. Curry, H. B. (1942). The inconsistency of certain formal logics. *The Journal of Symbolic Logic*, *7*(3), 115–117.
4. Goodship, L. (1996). On dialetheism. *Australasian Journal of Philosophy*, *74*(1), 153–161.
5. Grattan-Guinness, I. (1998). Structural similarity or structuralism? Comments on Priest's Analysis of the paradoxes of self-reference. *Mind*, *107*(428), 823–834.
6. Meyer, R. K., Routley, R., & Michael, D. J. (1979). Curry's paradox. *Analysis*, *39*(3), 124–128.
7. Pleitz, M. (2015). Curry's paradox and the Inclosure schema. In P. Arazim & D. Michal (Eds.), *Logica yearbook 2014*. London: College Publications.
8. Pleitz, M. ([forthcoming]): *Logic, language, and the Liar paradox*. Münster: Mentis.

[13] With a view of recovering the paradoxes, we could put the predicate-operator '$E(x : p)$' to the side, and work in its stead with an operator '$I : p$' with the intended meaning 'I now unambiguously say that p', governed by the principle $\forall p \forall q (I : p \wedge I : q \rightarrow (p \leftrightarrow q))$, that would here be justified already by the intended unambiguousness. Now we would get something like a Liar sentence and a Curry sentence by saying (!) '$I : \forall p (I : p \rightarrow \neg p)$' and '$I : \forall p (I : p \rightarrow (p \rightarrow q_0))$', respectively. However, the 'I say now that'-variant of Curry's paradox is likely to be less harmful because one can say only so many things, and thus it would be more difficult to trivialize the system.

9. Priest, G. (1994). The structure of the paradoxes of self reference. *Mind*, *103*(409), 25–34.
10. Priest, G. (2002). *Beyond the Limits of Thought* (2nd ed.). Oxford: Clarendon.
11. Priest, G. (2006). *In Contradiction*. Oxford: Oxford University Press.
12. Priest, G. (2014). What If? Part I: The Semantic Paradoxes. Draft of July 8, (2014[draft]).
13. Priest, G. (2014). *One. Being an investigation into the unity of reality and of its parts, including the singular object which is nothingness*, Oxford: Oxford University Press.
14. Prior, A. N. (1961). On a family of paradoxes. *Notre Dame Journal of Formal Logic*, *2*, 16–32.
15. Read, S. (2009). Plural signification and the Liar paradox. *Philosophical Studies*, *145*(3), 363–375.

Dialetheism in the Structure of Phenomenal Time

Corry Shores

Abstract In practice, phenomenology is an investigation of one's own consciousness by means of introspective awareness. Nonetheless, it can be considered a special sort of science, given that it obtains its data using a rigorous methodology. On the basis of these data, phenomenologists can devise "models" that describe the structures of consciousness. Husserl in fact thought that even the forms of logical judgments can be traced to more basic structures of consciousness. After examining the way that he locates the origin of negation in experiences of phenomenal "disappointment," which result in part from the layered structure of time-constituting consciousness, we turn to Barry Dainton's construction of models of the specious present. One type that is built upon Husserl's writings is a "retentional" model where the objective present is a simple instant, but all the while other recent moments have stacked up in retentional awareness to create the illusion of a present with a durational thickness. In Dainton's own rival "extensional" model, however, the present really does extend for a duration of about a second or so, and all the moments that seem present in fact are. At the end I propose a model of the specious present that is based on Graham Priest's spread hypothesis. It does not vindicate dialetheism; rather, it is merely built upon the assumption that we directly perceive dialetheias of motion. It is both retentional and extensional, since in it the actual present of our conscious activity has a very tiny extensive spread, all while recent prior spreads stack up in our retentional awareness to create the impression of an enduring present. This model has the advantages of explaining the continuity of phases in the specious present while also accounting for experiences of phenomenal disappointment.

Keywords Dialetheism · Specious present · Phenomenology · Edmund Husserl · Barry Dainton · Graham Priest

---

C. Shores (✉)
Ankara University, Ankara, Turkey
e-mail: corryshores@gmail.com

H. Andreas and P. Verdée (eds.), *Logical Studies of Paraconsistent Reasoning in Science and Mathematics*, Trends in Logic 45,
DOI 10.1007/978-3-319-40220-8_10

1 Introduction: Phenomenology as a Science

The founder of modern phenomenology, Edmund Husserl, considered his new field of inquiry to be a *science* like all other disciplines taking that name. This claim is controversial given that phenomenology obtains its evidences by means of subjective introspection rather than from objectively verifiable scientific experimentation. And also, the data obtained using the methods of phenomenology are the contents of our awareness. Yet, these contents are given to us in a ceaseless "Heracleitean flux," and for that reason they would seem to be unable to provide us with any self-consistent entities whose regularities can be scientifically determined [1, pp. 77–78]. However, Husserl invented reliable methodologies, called the "reductions," to study the contents of our consciousness. They allow us to be aware both of the continuously varying stream of contents and as well, to some degree, of the fixed fundamental structures of our consciousness. The eidetic reduction, for example, enables us to study the essential structures of a phenomenon by stripping away all of its accidental phenomenal determinations. Husserl considered phenomenology more specifically to be an a priori science, like mathematics for example. Such disciplines may begin as pure a priori sciences but can later come to find application in empirical studies, as for example how mathematics and geometry were for the most part developed independently of physics but were later successfully applied in this domain [1, p. 83]. Nonetheless, phenomenology is not purely a priori like math, since it studies actual experiences happening in the present, and yet it still is methodologically distinct from the empirical sciences. If we allow the name, we might call it a "subjective science." Using contemporary phenomenologist Barry Dainton's usage of the term "model," we may say that phenomenology is a science that produces models of the structures of consciousness. By "model" Dainton means simply a description, sometimes depicted with diagrams, of the structural features of consciousness and their relations. A general methodology that one may use in phenomenology is as follows. (1) By employing Husserl's reductions or some other method of introspective analysis of one's own acts of consciousness, one discovers specific phenomenal properties of these experiences. For example, in our temporal consciousness, time as a phenomenon is given with the phenomenal property of continuous flow. Then (2) one constructs models whose described structural elements should explain how it is that consciousness is constituted in such a way as to produce these phenomenal features. (3) The model is then tested to determine whether the phenomenal traits that it implies we should experience are ones that we really do find in our introspections, and also we look to see if there are still features of experience that the model should account for but fails to do so. This evaluation is performed by again introspectively analyzing one's own awareness, and it helps determine the viability of the model.[1]

[1]This specific sort of methodology is not shared by all phenomenologists, since Husserl for example was less concerned with the second and third steps. For him, phenomenology should be conducted in a mode of "understanding [*Verstehen*]" rather than in a mode of "explanation [*Erklärung*]" [7, pp. 1–14], and thus his work was not devoted to creating explanatory models. This other task is something that has been taken up more recently for example by phenomenologist

2 Husserl and the Phenomenology of Logic

To conduct this investigation into the logical principles underlying our models of time-consciousness, we should begin first by seeing how it is that phenomenological models can be formed on the basis of logical relations and principles. Such a study can for example be found in Edmund Husserl's *Experience and Judgment: Investigations in a Genealogy of Logic*. Here he argues that in order to attain a "comprehensive concept of logic," we need in addition to digging into its historical ancestry to as well probe into its phenomenological "genealogy." By this he means that in our studies of logic we want to know more than just the rules and principles of formulation, inference, and other formal concerns; we as well need to know in what way consciousness provides the conditions for logic judgments to be constructed in the first place.

> Now, if the logician really aims at a logic in the comprehensive and serious sense of the word, then his interest is directed toward the laws of formation of judgments – the principles and rules of formal logic – not toward the mere rules of a game but toward rules which the constitution of the forms must satisfy if any knowledge whatever is to be possible. [6, pp. 16–17]

Husserl takes particular care in this book to analyze predicative judgments, since they stand "at the *center of formal logic* as it has developed historically" [6, p. 11]; for, "[s]ince Aristotle, it has been held as certain that the basic schema of judgment is the *copulative* judgment, which is reducible to the basic form *S* is *p*" [6, p. 15]. As we dig deeper into the structures and processes of consciousness, we then discover

> not only that logical activity is already present at levels in which it was not recognized by the [logical] tradition and that, accordingly, the traditional logical problematic begins at a relatively higher level, but that, above all, it is precisely in these lower levels that the concealed presuppositions are to be found, on the basis of which the meaning and legitimacy of the higher-level self-evidences of the logician are first and ultimately intelligible. Only in this way will it be possible to come to grips with logical tradition in its entirety, and – as a further, distant goal of the phenomenological elucidation of logic – to attain that comprehensive concept of logic [...] of which we spoke. [6, p. 13]

So while *S is p* is a form of judgment conducted on a higher level of consciousness, it is made possible by means of structures on a lower "passive" level of object

(Footnote 1 continued)
Barry Dainton, whose models we discuss later and whose works inspire many of the basic questions and methodology of our treatment here. He draws from Husserl's descriptive analyses of time consciousness and formulates models of the "specious present," aiming to make them as faithful as possible to Husserl's descriptions. As we noted, Husserl generated these descriptions presumably by performing his "reductions," with another example being the *epoché* or "phenomenological reduction." By means of it, we may turn our awareness away from our normal consciousness of objects, in which we regard them as being existing things, and move our attention instead toward those conscious acts through which we experience their appearing and as well toward the structures of consciousness involved in these experiences [8, pp. 51–62]. Yet, Husserl's descriptions as we said do not go as far as presenting explanatory models, and so it is not always obvious what the best way is to formulate those models. See for example Gallagher's [5] and Dainton's [3] debate regarding the proper modelling of Husserl's structures of time-consciousness.

constitution. For, in order to judge some subject as having a certain predicate, our consciousness in a more basic way needs to be structured so that particular contents are phenomenally given not just as they are in themselves but as well as being attributed to some more generalized phenomenal component of our awareness.

The basic structure of consciousness that enables us to make predicative judgments, according to Husserl, is a "mental overlapping" of the apprehension of the subject with the apprehension of its predicate. In this way, consciousness takes the form of a "single double ray," meaning that in one synthetically unified act of present awareness we direct our attention in a doubled way to both the phenomenal substrate and to its "pre-predicated" determinations [6, p. 115]. This structure of doubled awareness is at work in all acts of consciousness; for, never are the data of our awareness somehow given to our consciousness as though they were not parts or features of something greater. There is an automatic, "passive," operation that synthetically groups contents such that no individual datum stands alone. In our visual perception, for example, we never see any color variation independently, but rather, regions of variation are automatically grouped together and appear as attributes of singular things.

So as a stream of phenomenal data is given to us in its continual variance, we all the while attribute these contents to one unified phenomenal object or another. This constant phenomenal elaboration of objects' properties Husserl calls "explication." He formulates the substantial unity of the thing being explicated and the phenomenal data that explicate it in the following way.

> Let us take an object, call it S, and its internal determinations $\alpha, \beta, \ldots$; [...] in the whole process of individual acts which lead from the apprehension of S to the apprehension of $\alpha, \beta, \ldots$ *we come to know* S. [...] Through the entire process the S retains the character of *theme*; and while, step by step, we gain possession of the moments, the parts, one after the other – and each one of them is precisely a moment or part, i.e., what is generally called a property or determination – each is nothing in itself but something of the object S, coming from it and in it. [...] In the development, the indeterminate theme S turns into the *substrate* of the properties which emerge, and they themselves are constituted in it as its *determinations*. [6, pp. 113–114]

So the reason we can form predicative judgments in the first place, Husserl argues, is because our consciousness contains a structure that allows the flux of determinate phenomenal contents to be synthetically co-apprehended with a more constant substrate-object, and by means of this synthesis the object is attributed with its determinate properties.

Husserl then makes the following distinction, which will lead us into our discussion of time consciousness. The two parts in the phenomenal structure of "explicative coincidence" that we just noted above are both given as simultaneous with one another. Yet, there is another sort of coincidence in the structures of consciousness that brings together contents given in different moments. We are not just aware of whatever is contained in any one instant, but in addition, our consciousness somehow reaches out beyond the present such that whatever has happened very recently is also quite apparent now. Barry Dainton calls this "diachronic co-consciousness" [2, p. 113]. By means of such a structure containing temporalized parts, Husserl argues, we are aware that from one moment to the next it is one same substrate *S* receiving all of its temporally varied determinations. This he calls the "total coincidence of identity," which is made possible by the structure of time-constituting consciousness [6, pp. 116–118].

3 Husserl and the Structure of Time-Constituting Consciousness

Time consciousness, in Husserl's analysis, is composed of a tripartite structure. (a) Our *intentional* awareness is directed toward contents given in present immediacy. (b) Those moments that have passed remain under the "gaze" of our *retentional* awareness. And (c) in a less explicit way, our attention is as well directed toward what we anticipate we will be intentionally aware of in the next moment, by means of our *protentional* awareness [6, pp. 107–108].

An important complicating feature of Husserl's model is that in a following moment, we are not retentionally aware merely of the prior moment's intentional contents, that is, of what was previously present to consciousness. In addition, we also hold in our current retentional awareness the just prior protentional contents. In other words, within each present moment we are doubly aware of what actually has come into our consciousness, and additionally, of what we anticipated, perhaps inaccurately, to be there. This combination is evinced in cases of surprise. According to this model, we are confused in these moments because what we anticipated to come into our awareness is regarded, on a higher level of consciousness, as being somehow contradictory to what actually did come into mind, even though we apprehend doubly these incompatible contents in one conjoining act of consciousness.

4 The Phenomenological Origins of Negation

In Husserl's account, normally our anticipations flow seamlessly into our forthcoming present intentions, and this continual matching of overlapping temporalized contents he calls "fulfillment" [6, p. 87]. However, in those moments of surprise and confusion when our prior anticipations do not match with our actual current experiences, we undergo phenomenal "disappointment," which he considers to be "the origin of negation" [6, p. 88]. For example,

> suppose that we have observed a ball uniformly red; for a time the course of the perception continues in such a way that this apprehension is *harmoniously* fulfilled. But now, in the progress of the perception, a part of the back side, not seen at first, is gradually revealed; and, in opposition to the original prescription, which ran "uniformly red, uniformly spherical," there emerges a consciousness of otherness which disappoints the anticipation: "*not* red, but green," "*not* spherical, but dented." ([6], p. 88, emphasis mine)

Using Husserl's previous notation, let us consider the ball's phenomenal substrate as S, the visual contents of its seemingly uniform redness as α, and its forthcoming additional greenness as β. While being aware of the substrate S, we first attribute α to S, then in the next moment $\sim\alpha\& \beta$ to S. But this does not yet take into account the disappointing non-fulfillment of our prior anticipation. If we only consider that the ball is not just red but also green, that does not alone account for why we are surprised. As we noted, in addition to seeing that the ball is $\sim \alpha$, we also keep just as vibrantly and potently in mind our retained prior expectation that it will be α, and thus the awareness of $\sim\alpha$ & β, in such cases of disappointment, is also $\alpha\& \sim\alpha$, or if expressed in the form of propositions: "the ball is uniformly red" and "it is not that the ball is uniformly red."

One important thing to note in this example is that the substrate Sremained intact despite it having inconsistent determinations. Yet, Husserl offers an example of phenomenal "doubt" where even the identity of the substrate comes to be in contradiction with itself for a sustained period.

> perhaps we see a figure standing in a store window, something which at first we take to be a real man, perhaps an employee working there. Then, however, we become hesitant and ask ourselves whether it is not just a mere mannequin. With closer observation, the doubt can be resolved in favor of one side or the other, but there can also be a period of hesitation during which there is doubt whether it is a man or a mannequin. In this way, *two perceptual apprehensions overlap* [...]. ([6], p. 92, emphasis mine)

In the prior case of the ball, there was "a radical break in the form of a decisive disappointment, thus [...] a conflict of an anticipatory intention with a newly emerging perceptual appearance, resulting in the cancellation of the first" [6, p. 92]. In that case, the contradiction held for no longer than a fleeting moment. Here, however, we begin with a stream of contents which leads us to anticipate the substrate for a man, and there comes a period when "there is superposed on it the sense 'clothed mannequin'" [6, p. 92]. During this period of uncertainty, there is a sustained "undecided conflict" where

> *Neither of the two is canceled out* [...]. *They stand in mutual conflict; each one has in a certain way its own force*, each is motivated, almost summoned, by the preceding perceptual situation and its intentional content. But demand is opposed to demand; one challenges the other, and vice versa. ([6], p. 92, emphasis mine)

In this case, to one same stream of determinations $\alpha, \beta, \ldots$ we are co-consciously attributing these contents to two inconsistent substrates, man and mannequin. Yet in these experiences, a phenomenal trait of this combination is not just its binarity but as well the tension they produce. That we are conscious of the figure being both a man and a mannequin is not enough to produce this phenomenal trait of tense conflict or opposition. What is needed as well, like in the prior ball example, is that we be conscious of negation, that is, "the figure is a man" and "it is not that the figure is a man." Both the one phenomenal sense and its negation are conjoined phenomenally by means of synthetically overlapping branches of our consciousness, and their prolonged inconsistency is what gives these experiences their phenomenal trait of irresolution.

Merleau-Ponty, a phenomenologist who takes up and elaborates Husserl's tripartite model of time consciousness, describes a situation that also exemplifies these states of phenomenal doubt.

> If I walk along a shore towards a ship which has run aground, and the funnel or masts merge into the forest bordering on the sand dune, there will be a moment when these details suddenly become part of the ship, and indissolubly fused with it. As I approached, [...] I merely felt that the look of the object was on the point of altering, that something was imminent in this tension, as a storm is imminent in storm clouds. [10, p. 20]

There is a sustained period when certain "trees" (which are really ship masts) in the hull's background seem both to be trees and seem to not be trees, perhaps because they do not sway in the wind in the same manner.

Note that both Husserl and Merleau-Ponty consider these experiences of phenomenal inconsistency to be exceptional cases. For Husserl, once one of the conflicting senses of the phenomenal object is judged to be the only valid one, the other is "nullified" and deemed invalid in our consciousness. Furthermore, Husserl says, our consciousness retroactively modifies all its retentions of the invalid one, "painting over" them with the newly validated object constitution. In other words, Husserl thinks that not only is the tension eventually dissolved into consistency, but all prior moments of this inconsistency are retroactively dissolved as well ([6, pp. 92–94], [9, pp. 68–71]). Similarly, Merleau-Ponty thinks that the phenomenal world is always in an "organic" state of intermeshed harmonious consistency. In the case of the ship masts, for example, to experience the figures for a brief time as trees and conjointly as not being trees is not really a case of true phenomenal inconsistency. This is because Merleau-Ponty assumes that even when we first began seeing the tree-like figures, we had implicitly constituted them as ship masts, yet this more valid sense lied unnoticed in the margins of our awareness. Thus always on some level, in Merleau-Ponty's model, phenomena are consistent with themselves, even if we are misled at times to not notice this explicitly [10, p. 20].

Nonetheless, might Merleau-Ponty and Husserl still be suggesting that given time-consciousness' overlapping structure, we could still understand those moments of phenomenal disappointment or doubt as being to some extent experiences of dialetheias, since both the phenomenon and its negation are conjoined and regarded as having equal validity? It seems not, since these exceptional experiences are not of dialetheias in the sense of true contradictions in the physical world. The figure in Husserl's example of doubt for some time is regarded both as a mannequin and not as one, but in reality it is only one of the two possibilities, and we eventually discern the correct one [6, p. 92]. Also, on the level of direct immediate perception, the ball never had contradictory perceptual determinations. This was only the case on a higher conscious level of "apperception," which is aware of more than just the immediately given contents [9, pp. 624–626]. Thus, the ball was not really at the same time entirely red and also not entirely red. Rather, the features of its other side were merely anticipated incorrectly at first. Such "dialetheias," then, seem really to be phenomenally artificial and anomalous.

5 The Specious Present

In order to work toward a model of time consciousness based on an awareness of actual dialetheias in the perceived physical world, we will look first at ways of modeling what is often called "the specious present." The present as we experience it does not seem to be an instantaneous snapshot giving us only the singularly present contents, nor are we aware only of this moment and the immediately prior one. In fact, we seem to be directly aware of very many instants in a series that all appear to be equally present. When seeing a moving object, for example, it is as if many successive positions, and thus many successive moments of our consciousness, are all "hanging" in our present experience. We are aware, then, of a little window of present time and not just the tiniest sliver. Graham Priest offers the following illustration for this phenomenal experience.

> A graphic way of focusing attention on the extended present is by concentrating on our experience of certain sorts of motion. For example, consider an analog watch or clock, with hands for the hour, the minute, and the second. One cannot see the minute hand (and a fortiori the hour hand) move— unless it is of the kind that jumps occasionally. One sees it in a certain position and infers that it has moved, since one remembers its being elsewhere. The second hand, by contrast, can actually be seen to move. One does not infer its motion by comparing present position with remembered position. Its motion is part of the phenomenological furniture. It is as if one can see the whole of a short stretch of motion at once. But of course, every point of the motion occurs at a different instantaneous time. The conclusion that we experience a present extended through a certain period of time seems mandatory. [11, p. 217]

This phenomenal feature was termed the "specious" present, since it was assumed to be an illusion: the vibrancy of retained just-passed contents trick us into experiencing them as present when in fact the present could have no such extended duration. Contemporary phenomenologist Barry Dainton has thoroughly analyzed the historical development of phenomenological models of the specious present, identifying problems in each, and on the basis of these analyses, proposes his own original "overlap model." It structures the specious present as objectively extending some brief period, somewhere in the neighborhood of a second or so [2, p. 113].[2] The alternate sort of model is the retentional one, like Husserl's, which structures the objective present as only an instant long, while the vibrancy of retentions gives our present experience an *illusory* "thickness." For a variety of reasons that will not concern us here, Dainton concludes that retentional models are not viable and that his proposed extensional overlap model succeeds where they fail.

---

[2] Dainton writes that there are no scientific studies which directly address the question of how long the specious present lasts. He does cite ones that indicate that the contents of our experience hold together in units of about three seconds long. Yet, these studies do not determine whether or not all those contents are perceived as present or if they also include memorial content. Without adequate scientific data to determine the length of the specious present, Dainton says we must use our own introspection. He reports that his specious present seems to last for a half of a second or so, but he generally uses the approximation of around one second [2, pp. 170–171].

In the overlap model, the specious present is one singular, but flowing, act of consciousness. Yet, it is long and complex enough that it can be said to contain a series of distinct phases or moments that pass into one another within that singular present. Nonetheless, all of them somehow are equally present to one another, both objectively and phenomenally. To explain the continuous flow from one specious present to another, Dainton's model depicts that movement as being based on an overlapping structure such that part of the prior act of consciousness carries over into the current one. But, since none of the present contents are retentional, that overlap cannot be the superposition of a retended memorial present with an actual immediate present. Instead, in his model one present overlaps by *sharing common parts* with the prior present. Part of the prior present somehow hangs in the current present, and it "overlaps" in the sense that it continues on in it. He distinguishes these two structures with diagrams similar to the ones below (Fig. 1).

So, how would Dainton's extensive overlap model explain the experience of temporal inconsistencies that Husserl described? As we see from the diagram, within the specious present, it appears that no content could be overlapping in a manner that would allow for self-contradiction (Fig. 2).

Thus, although Dainton's model does account for the phenomenal trait of the present's durational quality, it is not entirely clear from it how we would experience phenomenal disappointment, since the contents from one present to the next are the self-same content and hence would not be held in consciousness in a conjunctive structure. Yet our concern is not just in formulating a model that explains such phenomenally artificial "dialetheias," but as well it will assume that we experience actual dialetheias existing in the changing physical world that we perceive. To see how such a model could be constructed, we turn now to Graham Priest's "spread hypothesis."

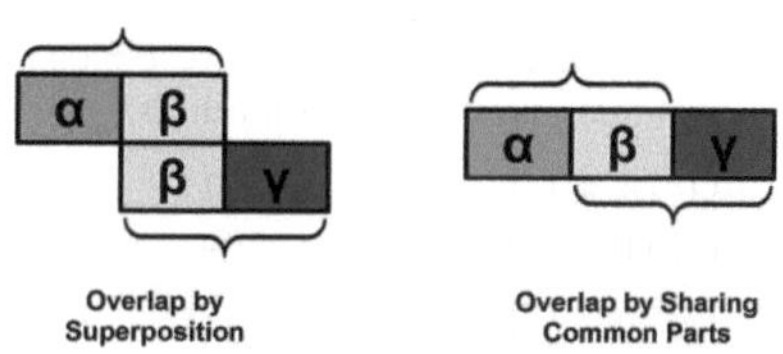

Fig. 1 Modified versions of Barry Dainton's diagrams distinguishing the two types of structural overlap in temporal consciousness [3, p. 21]

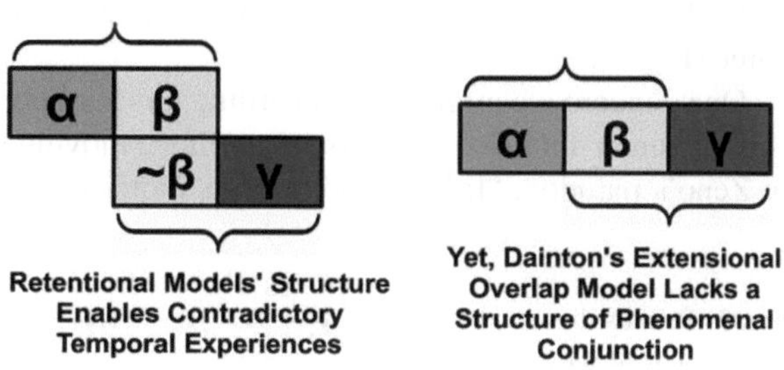

Fig. 2 It is not clear how Dainton's model would account for the experience of temporally contradicting contents

6 Graham Priest's Spread Hypothesis Applied to the Specious Present

In his *In Contradiction*, Graham Priest notes how a conception of the specious present as being extensive can lead to absurdity, even though experience tells us that this is really so. For, "[h]ow can we possibly experience two times at the same time? By the time we experience the later one, the earlier one must be over" [11, p. 217].

However, he says that "the extended present is accommodated very happily by the assumption that time itself satisfies the spread hypothesis" [11, p. 217]. He explains his spread hypothesis earlier in this book in the context of accounts of physical motion. We begin first by considering the "Russellean" or "orthodox" account. According to Bertrand Russell, an object is in motion merely because it is *at* some location *at* some instant, and *at* some other location *at* some other instant, and it is *at* intermediary positions in between [12, p. 84]. However, bodies can never be in more than one place at the same time [13, p. 473], and thus the object is never in a state of transitional motion from place to place.

> *we must entirely reject the notion of a state of motion.* Motion consists merely in the occupation of different places at different times [...]. *There is no transition from place to place*, no consecutive moment or consecutive position, no such thing as velocity [...]. ([13], p. 480, emphasis mine)

With this in mind, Priest formulates the "Russellean state description" of motion. Consider an object whose various positions at certain times are described by some function. In the orthodox account, at some given point in time, the object is only at the location determined by the function, and not anywhere else.

> Now, consider a body, b, in motion [...] moving along a one dimensional continuum, also represented by the real line. Let us write Bx for 'b is at point x'. Let us also suppose that each real, r, has a name, $\underline{r}$. [...] Let the motion of b be represented by the equation $x = f(t)$. Then the evaluation, v, which corresponds to this motion according to the Russellean account, is just that given by the conditions:
>
> (1a) $1 \in v_t(B\underline{r})$ iff $r = f(t)$
>
> (1b) $0 \in v_t(B\underline{r})$ iff $r \neq f(t)$
>
> [11, p. 177]

In Priest's diagram, we see how only one position is assigned to only one time value (Fig. 3).

One concern we might raise regarding this Russellean state description of motion is its "counter-intuitiveness;" for, regarding a moving body, as for example the arrow in Zeno's paradox, "[a]t any point in its motion it advances not at all. Yet in some

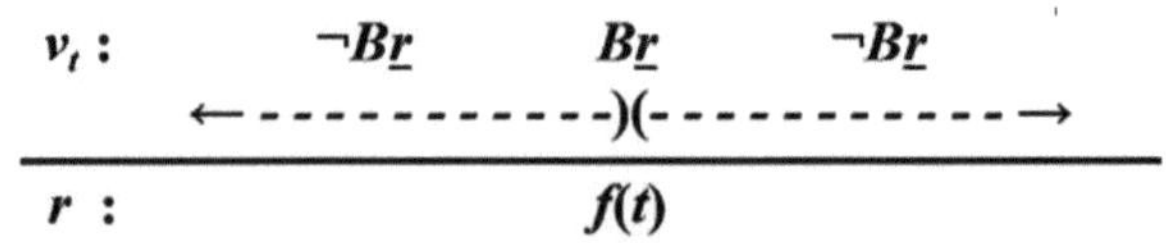

Fig. 3 Priest's diagram of the Russellean state description (modified slightly from [11, p. 178])

apparently magical way, in a collection of these it advances. Now a sum of nothings, even infinitely many nothings, is nothing. So how does it do it?" [11, p. 175; 180].

Opposed to this is the "Hegelian" account of motion. Priest quotes the following from Hegel's *Science of Logic*: "motion itself is contradiction's immediate existence. Something moves not because at one moment of time it is here and at another there, but because at one and the same moment it is here and not here . . ." (Hegel, qtd. in [11], p. 175). In this Hegelean conception, the location of moving objects cannot be localized during very tiny intervals of time [11, p. 176]. With this in mind, Priest formulates his *spread hypothesis*: "A body cannot be localised to a point it is occupying at an instant of time, but only to those points it occupies in a small neighbourhood of that time" [11, p. 177]. He incorporates his spread hypothesis into his formulation for the "Hegelean state description" of motion. In this case, at some specific point in time during the object's motion, there would be a tiny spread of neighboring time points around it during which the object would be found at all points within a tiny spread of space.

> In accordance with the [spread] hypothesis, there is an interval containing t, θ_t [...] such that, in some sense, if $t' \in \theta_{t\ \upsilon}$, b's occupation of its location at t' is reproduced at t. I suggest that a plausible formal interpretation of this is that the state description of b at t is just the "superposition" of all the Russellean state descriptions, $\upsilon_{t'}$, where $t' \in \theta_t$. More precisely, it is the evaluation, υ, given by the conditions
>
> (2a) $1 \in \upsilon_t(B\underline{r})$ iff, for some $t' \in \theta_t$, $r = f(t')$
>
> (2b) $0 \in \upsilon_t(B\underline{r})$ iff, for some $t' \in \theta_t$, $r \neq f(t')$
>
> [11, p. 178]

We then "write Σ_t for the *spread* of all the points occupied at t" [11, p. 178]. In the diagram below, we see depicted the small spread of time points in Σ_t, during which the object will be found at all locations within the corresponding spread of spatial points (Fig. 4). So long as the spread of time points around t really do go slightly beyond it, then "at t a number of contradictions are realised. For all $r \in \Sigma_t$, $1 \in \upsilon_t(B\underline{r} \wedge \neg B\underline{r})$" [11, p. 178].

Returning to the section of Priest's book on the specious present, we may see what he means when writing that the spread hypothesis can accommodate the "extended" (specious) present [11, p. 217]. Here we are applying the spread hypothesis to just time itself rather than to motion. In such a temporal application we would for example say that "at 12 noon it is every time around 12 noon" [11, p. 215]. Similarly, we would say that at any given moment of our consciousness, we are aware also of every moment

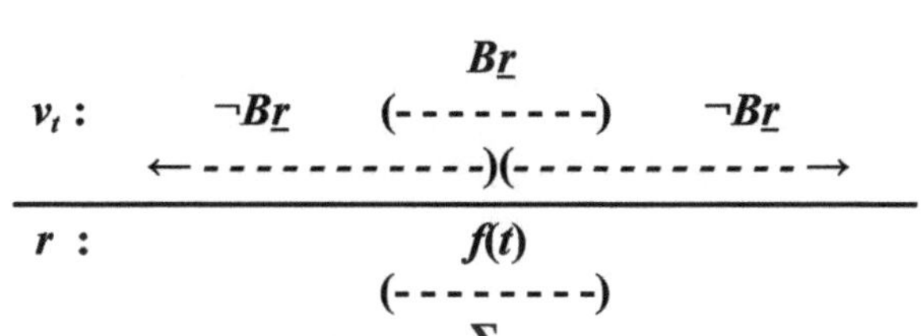

Fig. 4 Priest's diagram of the Hegelean state description (modified slightly from [11, p. 178])

around that instant [11, p. 217]. Thus "[t]here is, to put it picturesquely, some past occurring at the present. The extended present just is the spread of time around the present (or perhaps just some part of it if we do not experience it all)" [11, p. 217]. As well, it would seem then from this formulation that at some given moment of awareness, we are equally aware of the contradicting contents in the nearby times. We now ask, how would the spread hypothesis be implemented in a model of the sort that Dainton constructs?

7 Conclusion: Modeling the Specious Present with Priest's Spread Hypothesis

Like Priest, Dainton also finds the Russellean at-at account to be counterintuitive, but for Dainton its problems are also phenomenologically evident. "Intuitively it seems wrong," he says, since "there seems to be a big difference between something that is flashing through the air and something that is resolutely motionless." Common sense, he continues, tells us that we directly see movement itself, because "objects in motion look different than their static counterparts. […] A sequence of static snapshots looks different than things that move" [4]. In Dainton's extensional model, our direct awareness of transitional movements is our consciousness of the content within one of the specious present's smallest phases "giving way to" and "flowing into" the content of the next tiny phase within that same specious present [2, p. 173]. Thus, at each smallest phase of the specious present, we are directly aware of moving objects' states of transition and not just of their fixed positions. But, it is not clear if Dainton thinks that momentary transitions in the physical world involve dialetheias where the object both is and is not in a certain location at a certain time. Yet, before we design a model based on the direct perception of real dialetheias in the physical world of change, we should note another element of Dainton's model that we might want to modify. The extensive "spread" of the specious present for him endures for about a half of a second to a second or so, rather than being a matter of very close neighboring instants. Yet, perhaps for one reason or another we would not want our model to assume that the present moment in the physical world has such a relatively long duration. The dialetheic model of the specious present that I propose would be both retentional and extensional, even though in Dainton's classifications, the models are only one or the other type. In ours, there is an objective present with a tiny spread of neighboring instants during which we are directly aware of dialetheias of motion. So in our visual experiences of moving objects, we would each moment see a tiny blur rather than a perfectly resolute still image. All the while, we still hold in our retentional consciousness many prior tiny blurs, which all blend together in our visual awareness as the blurry streak we sometimes see when viewing objects speeding by.

This model, then, does not vindicate dialetheism, but rather presupposes it. Nonetheless, as a phenomenological model, it has certain advantages, namely, the retentional overlapping of its tiny spreads does well to explain our experience of the blended continuity of phases in the specious present while also providing a basis to account for the experiences of disappointment that Husserl describes. A dialetheist, then, has this model as one option for modelling the specious present in a phenomenologically useful way. This model views the sequence of moments contained in the specious present not as a series of mere contents but instead as a series of contradictory ones, the combination of which in our retentional awareness giving us the impression that at present a brief slice of change and temporal passage is taking place. This could serve as one way that phenomenological studies of time may benefit from incorporating the recent advances in paraconsistent logic in efforts to create new models of time-constituting consciousness. Dialetheias have been uncovered both in the physical world, in states of motion for example, and in the workings of language, evinced for instance in the liar's paradox. To these we could add that dialetheias might be found at the basis of every moment in our temporal experience.

Acknowledgments The author would like to thank Ullrich Melle of the Husserl Archives at the University of Leuven for contributing helpful comments and advice, and as well he thanks the anonymous referees for their corrections and recommendations.

References

1. Bernet, R., Kern, I., & Marbach, E. (1993). *An introduction to Husserlian phenomenology*. Evanston: Northwestern University.
2. Dainton, B. (2000). *Stream of consciousness: unity and continuity in conscious experience*. London: Routledge.
3. Dainton, B. (2003). Time in experience: reply to Gallagher. *Psyche*, *9*(12)
4. Dainton, B. (2011). Passage in Experience and Reality. Lecture, *Cosmological and Conscious Time* at the Van Leer Jerusalem Institute, Jerusalem, Israel, 28 June 2011.
5. Gallagher, S. (2003). Sync-ing in the stream of experience: time-consciousness in Broad, Husserl, and Dainton. *Psyche*, *9*(10)
6. Husserl, E. (1973). *Experience and judgment: investigations in a genealogy of logic*, L. Landgrebe (Ed.), trans. James S. Churchill and Karl Ameriks. London: Routledge & Kegan Paul.
7. Husserl, E. (1977). *Phenomenological psychology, lectures, summer semester, 1925*, trans, John Scanlon. Nijhoff: The Hague.
8. Husserl, E. (1983). *Ideas pertaining to a pure phenomenology and to a phenomenological philosophy. First Book. General introduction to a pure phenomenology*, trans. F. Kersten. The Hague: Nijhoff.
9. Husserl, E. (2001). *Analyses concerning active and passive synthesis: lectures on transcendental logic*, R. Bernet (Ed.), trans. Anthony J. Steinbock. Dordrecht: Kluwer.
10. Merleau-Ponty, M. (1958). *Phenomenology of perception*, trans. Colin Smith. London: Routledge.
11. Priest, G. (2006) [1987]. *In contradiction: a study of the transconsistent.* 2nd Edn. Oxford: Oxford University.
12. Russell, B. (1917). [1910]. Mathematics and the metaphysicians. In *Mysticism and logic, and other essays*, 74–96. London: George Allen & Unwin.
13. Russell, B. (2010). [1903]. *Principles of mathematics*. London: Routledge.

Saving Proof from Paradox: Gödel's Paradox and the Inconsistency of Informal Mathematics

Fenner Stanley Tanswell

Abstract In this paper I shall consider two related avenues of argument that have been used to make the case for the inconsistency of mathematics: firstly, Gödel's paradox which leads to a contradiction within mathematics and, secondly, the incompatibility of completeness and consistency established by Gödel's incompleteness theorems. By bringing in considerations from the philosophy of mathematical practice on informal proofs, I suggest that we should add to the two axes of completeness and consistency a third axis of *formality* and *informality*. I use this perspective to respond to the arguments for the inconsistency of mathematics made by Beall and Priest, presenting problems with the assumptions needed concerning formalisation, the unity of informal mathematics and the relation between the formal and informal.

1 Introduction

Is mathematics consistent? While in practice we generally proceed as if it is, for dialetheists such as Priest in [15], mathematics is one of the main battlegrounds on which to establish that inconsistencies do indeed arise and require their dialetheist solutions. In this paper I shall consider two related avenues of argument that have been used to make the case for the inconsistency of mathematics: firstly, paradoxes which lead to contradictions internal to mathematics and, secondly, the incompatibility

This work was supported by the Caroline Elder Scholarship and a St Andrews/Stirling Philosophy Scholarship. Many thanks to two anonymous referees, Aaron Cotnoir, Benedikt Löwe, Noah Friedman-Biglin, Jc Beall, Graham Priest, Alex Yates, Ryo Ito, Morgan Thomas, Brian King and audiences in St Andrews, Cambridge and Munich.

F.S. Tanswell (✉)
Department of Philosophy, Arché Philosophical Research Centre,
University of St Andrews, St Andrews KY16 9AL, St Andrews, UK
e-mail: ft34@st-andrews.ac.uk

F.S. Tanswell
Department of Philosophy, University of Stirling, Stirling FK9 4AL, Stirling, UK

H. Andreas and P. Verdée (eds.), *Logical Studies of Paraconsistent Reasoning in Science and Mathematics*, Trends in Logic 45,
DOI 10.1007/978-3-319-40220-8_11

of completeness and consistency established by Gödel's incompleteness theorems. These two strands of argument are closely connected, for the most apparently problematic paradox in the case of mathematics is *Gödel's paradox*, that of the sentence which says of itself that it is unprovable, which is closely related to common constructions of Gödel sentences for formal systems whereby we get to the balancing act between completeness and consistency.

My response to the two lines of dialetheist argument will bring in considerations from the philosophy of mathematical practice on the nature of informal proofs. One thing I will argue for is that we should add to the two axes of completeness and consistency a third axis of *formality* and *informality*. Given this third axis, we can consider the dialetheist arguments in two different ways. At the informal end, the previously problematic paradoxes may be genuine, but I argue that there is no compelling reason to see them as internal to mathematics. Meanwhile, at the formal end of the scale, considerations of the practical role of formalisation in mathematics will allow me to make a positive case for incompleteness over inconsistency without begging the question against the dialetheists. My main conclusion will be that the dialetheist arguments considered do not establish that mathematics is inconsistent.

Answering the ultimate question of whether mathematics is consistent from this perspective which encompasses informal proofs and mathematical practice would, I believe, be a major undertaking, and one which I am not intending to complete here. The intention is rather to take the first step in this direction by demonstrating that the matter is not already settled, since the standard arguments from Gödel's theorems and the paradox of provability do not succeed. In fact, I believe these arguments fall apart through a number of the assumptions they need about informal proofs, the nature of mathematics and the process of formalisation, so I shall proceed to raise these objections in turn.

To begin, Sect. 2 will introduce the key distinction between formal and informal proofs that my arguments will focus on. Next, in Sect. 3 I will lay out what Gödel's paradox is and why I do not take it to be a concern for mathematics. In Sect. 4 I lay out Priest's longer argument for the inconsistency of informal mathematics based on the application of Gödel's first incompleteness theorem to informal mathematics and the conclusions he draws from this concerning the *inherent* inconsistency of informal mathematics. In Sect. 5, I argue that the way of understanding formalisation on which Priest's argument succeeds is a bad one, then show that a better understanding means the argument no longer goes through. In Sects. 6 and 7, I argue against the thought that we can formalise mathematics as single theory, proposing that a better thought would be to approach formalisation in a fragmented way. Finally, in Sect. 8 I consider formality and informality as a third axis, and a final argument against Priest that he changes the subject in switching between the formal and informal.

2 Formal and Informal Proofs

Before we can begin, we need to be sufficiently clear on the distinction between formal and informal proofs, as this will play a central role in the remainder of this paper.[1]

Formal proofs are those which are studied in logic and proof theory, and may be defined in the usual way. For example, we might define a formal language, give rules for well-formed formulae in that language, specify axioms to be taken as basic and lay down inference rules for stepping between formulae. A *formal proof* (relative to such a specified system) will be a (usually finite) sequence of formulae where each is either an axiom or follows stepwise from previous formulae by an application of one of the inference rules, where the final formula is a statement of what was to be proven and is thus established as a theorem in the system.

However, formal proofs are rarely seen in actual mathematical practice. Instead the type of proofs that are employed by mathematicians in their daily activities, teaching and published work tend to be very different. In most cases no formal language is specified, axioms are rarely given and inferences are not confined to just the basic rules. Steps in these proofs can rather be leaps and invoke the background knowledge of your target audience, the semantic understanding of the terms being employed, visualisation, diagrams and topic-specific styles of reasoning. Let us call proofs in this sense *informal proofs*. Although this would be extremely unsatisfying as a definition, it is certainly not intended as such as one of the main challenges for philosophers of mathematical practice is to pin down exactly what counts as a good, legitimate, correct and rigorous informal proof and filling this out further would take me beyond the scope of this paper. Nonetheless, there is a good deal of literature that does deal with this issue that elaborates on the distinction I am invoking (see [8–11, 17, 19] etc.).

A number of the differences between these two types of proof will affect the assessment of whether the arguments I am considering successfully establish that mathematics is inconsistent. Gödel's first incompleteness theorem relates to proof as an explicitly defined, formal notion attached to a formal system and one of my main counter-arguments in what is to come is that this will not transpose across to apply to informal proofs. Gödel's proof tells us about the limits of formal systems *which meet certain conditions*, like having a certain amount of expressive power, being able to prove a certain amount of basic mathematics (enough to allow for the required coding etc.) and having an effective procedure for enumerating its theorems. What will be required for the dialetheist line to work, then, will be to show that informal proofs are close enough to formal ones to even begin applying these conditions and that they then meet them. I will argue to the contrary that informal proofs are sufficiently different that the proof will not apply. Some key differences of informal proofs that will play a role later include the social and contextual components of

[1] A terminological note: while I speak of 'informal proofs' and 'formal proofs', some of the literature on this subject instead speaks of 'proofs' and 'derivations' to get at the same distinction. In [15], Priest also uses the term 'naïve proof' to refer to the informal proofs.

whether such a proof is successful or not; the partially-fragmented nature of modern (informal) mathematics; and the fact that informal mathematics extends to include diagrammatic proofs which have more intuitive inferential rules. Finally, even if the dialetheist arguments manage to establish that informal proofs can be formalised appropriately, there will still be the need to show that the conditions are met.[2]

Before getting the details of the argument from Gödel's theorems, let me assess whether a simpler argument from paradox outlined by Beall is sufficient to show that mathematics is inconsistent.

3 Gödel's Paradox and Beall's Argument

The first argument I will consider comes from *Gödel's paradox*.[3] Let us begin, therefore, by examining the paradox:

> **GP**: This sentence is (informally) unprovable.

Suppose GP is false; then it is informally provable. Since we take our informal mathematical proofs to establish mathematical truths, it follows that GP is also true. Yet this contradicts the assumption that GP is false, so using proof by contradiction we establish that GP is true. However, since we have just proved GP, it is informally provable. But GP states that it is unprovable, so it must be false. Contradiction.

Now consider how it is that this paradox might show that mathematics is inconsistent. Beall gives the following argument:

> There seems to be little hope of denying that [GP] is indeed a sentence of our informal mathematics. Accordingly, the only way to avoid the above result is to revert to formalising away the inconsistency— a response familiar from the histories of naïve set theory, naïve semantic theory, and so on. If one does this, however, then (by familiar results) one loses completeness, which can be regained only by endorsing inconsistency. Either way, then, we seem to be led to inconsistent mathematics. [3, p. 324]

Setting aside the option to formalise away the inconsistency until Sect. 4, the initial argument is that since GP is part of mathematics and GP leads to an inconsistency, it must therefore be that there is an inconsistency in mathematics. In the rest of this section I will undertake the (purportedly hopeless) task of denying that GP is part of mathematics.

The only sensible suggestion as to why GP should be part of mathematics, it would seem, is that GP concerns the broadly mathematical concept of informal provability. I contend, though, that this is not sufficient to make GP a statement of mathematics. The reason is that I take the concept of informal proof to be used to talk and reason

---

[2]In [15], Priest argues that these conditions will be met. I believe that the flawed step in the argument is the earlier one of formalisation (as will be covered in Sect. 5), so I will not actively engage in a discussion about whether this formalisation will have an effective calculus etc.

[3]At this point we are only concerned with the informal version of the paradox. Later I take on the formal results of Gödel's theorems.

about mathematics without it being *a part of* mathematics. While the former is obvious, for the paradox to render mathematics inconsistent we actually need the later, more contentious claim. Of course, I hold that informal proof and provability are very important notions in talking about mathematics, but it is crucial to emphasise that these are notions *about* mathematics. To establish that the paradox will render mathematics inconsistent, though, we need the extra claim that it is *a part of* informal mathematics. In general, a statement being about mathematics and a statement being part of mathematics can coincide, but certainly don't always. Consider the following:

(1) Mathematics is traditionally done on blackboards.
(2) This square building with 12 m sides must have an area of $144\,\text{m}^2$.
(3) $111, 111, 111 \times 111, 111, 111 = 12, 345, 678, 987, 654, 321$.
(4) Ron likes bacon and eggs.

Here (1) is a statement about mathematics but is not itself a part of mathematics. In contrast, (2) is a mathematical statement which is being applied to a situation, so in a relevant sense is not about mathematics. The third item is both mathematical as a statement and about a mathematical fact, while the fourth sentence is neither. Since these two notions can be pulled apart with minimal effort, that a sentence falls under one of them certainly can't constitute a reason to think that it falls under the other. It can therefore be concluded that the notion of informal provability being about mathematics is not sufficient to establish that GP falls within mathematics.

One can also give positive arguments as to why informal provability should not be considered a concept within mathematics. For example, the lack of a precise mathematical definition we observed in Sect. 2 clearly supports the claim that informal provability is not a notion internal to mathematics. Nor does it interrelate with other mathematical concepts in the way that standard mathematical concepts do (such as, for example, group, integer, derivative, line, etc.). The only notable conceptual link it has is with truth, as exploited by the paradox, but if anything the informal notion of truth in mathematics (before being formalised into some formal theory of truth) will belong to the same category of notions about mathematics that are not within mathematics.

By denying that informal provability is a concept within informal mathematics, it can consequently also be denied that GP is a sentence of our informal mathematics. It is thus reasonable to deny that Beall has showed that informal mathematics is inconsistent by using GP. This certainly does not provide an ultimate solution to Gödel's paradox, but it does keep the derived inconsistency out of mathematics and allows us to set aside the paradox to be solved in line with whatever one's favourite solution is to paradoxes generally.[4]

Now, let me note two things about what has gone on here which will be recurrent throughout the paper. Firstly, although this section does not solve Gödel's paradox,

[4]A final note on Beall: although the argument I am criticising is from an older paper, the response offered here would fit well with Beall's more recent work in Beal [4]. The suggestion I have made may be appropriated to make the case that informal proof should join truth in the category of useful devices, which when introduced bring 'merely' semantic paradoxes as by-products or 'spandrels' without thereby rendering the base language (in this case, that of mathematics) inconsistent.

this is not really necessary for the purposes of the current project. Beall, Priest and others have a substantial case for the inconsistency of natural languages, a case which is not the target of this paper and would have to be addressed separately if one were so inclined. For both of these authors the claim that mathematics is inconsistent is an additional one that is supported by additional argumentation and it is precisely these arguments which I am targeting. Thus, by rejecting that Gödel's paradox is part of mathematics, what has been done is to show that these additional arguments do not cover more ground than the original case for the inconsistency of natural languages and therefore don't provide added support for dialetheism from the realm of mathematics. Secondly, the separation between being part of mathematics and the concepts used about mathematics is not just a way to re-introduce the object language/meta-language distinction for informal mathematics. A separation of languages is not important because the point is not really one about languages, instead it is about the subject-matter of mathematics. While we may use GP to argue that the concept of informal provability is inconsistent, this does no more work than the liar or any other semantic paradox unless it infects the realm of mathematics. As such, showing that informal proof is not the kind of thing to be investigated mathematically blocks the argument considered in this section.

4 Priest's Argument for the Inconsistency of Informal Mathematics

In Chap. 3 of [15], entitled "Gödel's Theorem", Priest makes use of Gödel's paradox in the same way as Beall subsequently went on to do, arguing that it shows that informal mathematics is inconsistent. In Priest's case, however, it is given as the culmination of a longer argument which aims to show that informal proof satisfies the conditions for Gödel's first incompleteness theorem in such a way as to lead to its inconsistency. This section will focus on explaining the details of Priest's argument.

Priest wants to show that informal proof is susceptible to Gödel's first incompleteness theorem. The first hurdle is that the theory of informal proofs is, on the surface at least, not formal and hence not immediately susceptible to Gödel's theorem. Priest addresses this in the following way:

> It should be said at once that naive proof, or at least the naive theory it generates, is not a formal theory in the sense of the theorem; but it is accepted by mathematicians that informal mathematics could be formalised if there were ever a point to doing so, and the belief seems quite legitimate. The language of naive proof, a fragment of English, could have its syntax tidied up so that it was a formal language, and the set of naïve theorems expressed in this language would be deductively closed. Hence we may, without injustice, talk about the naive theory as if it were a formal theory. [15, p. 41][5]

[5]As the target of his argument, Priest needs to explain what he takes naïve or informal mathematics to be exactly. He says:

In Sect. 5, I will claim that Priest's reasoning fails to go through at this point. For now, though, let us complete Priest's argument that informal proof satisfies the conditions of Gödel's theorem. The other pieces that Priest needs are that the formalised theory can express all recursive functions and that the proof relation of the formalised theory is recursive. He rightly takes the first requisite to be obviously satisfied and the second to be the contentious one, listing a number of possible objections and his replies. A discussion of these would be irrelevant to the purposes of this paper, so for now we shall grant that the formalised proof relation is recursive.

Given that Priest has now established that informal proof satisfies the conditions of Gödel's theorem, the thrust of his argument is as follows:

> For let T be (the formalisation of) our naive proof procedures. Then, since T satisfies the conditions of Gödel's theorem, if T is consistent there is a sentence φ which is not provable in T, but which we can establish as true by a naive proof, and hence is provable in T. The only way out of the problem, other than to accept the contradiction, and thus dialetheism anyway, is to accept the inconsistency of naive proof. So we are forced to admit that our naive proof procedures are inconsistent. But our naive proof procedures just are those methods of deductive argument by which things are established as true. It follows that some contradictions are true; that is, dialetheism is correct. [15, p. 44]

Priest soon makes the link between φ and Gödel's paradox. For if we take φ to be the formalisation of GP,[6] the inconsistency of Sect. 3 will quickly re-emerge within the formalisation of informal mathematics. A key point is that a standard move towards incompleteness over inconsistency is to separate the object language from the meta-language, but that here we are dealing with informal proof and informal mathematics, for which there is no such distinction, meaning that the orthodox move towards incompleteness is not available. Indeed, this is the entire point of focusing on informal mathematics.

The conclusion that Priest draws is that we are left with true contradictions and dialetheism.[7] Informal mathematics is seen to be inconsistent, but even more penetratingly he can claim that there is no escape from this application of the incompleteness theorems to informal mathematics and so "... we might say that our naive proof procedures are not just contingently inconsistent, but essentially so... [D]ialetheism

---

> Proof, as understood by mathematicians (not logicians), is that process of deductive argumentation by which we establish certain mathematical claims to be true. [15, p. 40]

His distinction is, in effect, the same as the distinction between formal and informal mathematics as found in Sect. 2.

[6]The matter is somewhat more complicated than this suggests, of course. Milne discusses in [12] the many ways that Gödel sentences can be constructed and what exactly they 'say'.

[7]Not just this, though, since Priest takes it that the theory given by the formalisation of informal mathematics can prove its own soundness and hence must be able to give its own semantics. From here he takes it to follow that it must be able to prove the T-scheme for this theory inside the theory, giving him all of the paradoxes he describes as semantic (as opposed to set-theoretic paradoxes). For example, he lists the liar, Grelling's paradox, Berry's paradox, Richard's paradox and Koenig's paradox as falling under the umbrella of semantic paradoxes. In fact, then, Priest argues that "Our naive theory is semantically closed and inconsistent. By contrast, any consistent theory cannot be semantically closed." [15, p. 47].

is inherent in thought." [15, pp. 47–48] That dialetheism is inherent in thought is one of the main claims of *In Contradiction*, supported by several pillars of argument. The argument described here that informal mathematics is *essentially* inconsistent forms one of these pillars, but I shall argue that this pillar will not hold any weight.

5 Formalising Mathematics

The move from the informal version of mathematics to a formalisation thereof is, in my opinion, too quick. By endorsing the claim that mathematicians take it that informal mathematics can be formalised, Priest moves from the informal theory to the formal one without much consideration of what this move entails or how the mathematicians he is invoking conceive of the formalisation process. For one thing, Priest might not want to endorse the naïve claims of mathematicians at all, since they most likely take mathematics to also be consistent. If such claims were definitive it might thus spell the end of dialetheism.

Nevertheless, it is worth considering how exactly the idea that mathematics should be formalisable will work precisely. In the first half of this section I discuss two options, along with how they interact with Priest's argument. The first follows a straightforward interpretation of Priest's claim but is shown to fail as an account of the formalisation of informal mathematics. The second avoids the problems with the first but, I argue, no longer lets Priest's argument go trough.

5.1 A First Option

Let us call the first option *many-one formalisation*.[8] The idea is that one takes the entirety of informal mathematics and tidies up the fragment of natural language expressing it to give a formal language. All of the informal theorems will have particular formal counterparts expressed in this one formal language, and the set of these formalised theorems is then deductively closed. For the first option, we consider this as the one single correct formal counterpart for the informal mathematics, a type of super-theory[9] of mathematics, in which all the current basic assumptions and their consequences are contained. This mirrors a standard idea of formalisation involving a routine procedure of 'filling in the gaps' (as is discussed, for instance, in the debate between Rav [17, 18] and Azzouni [1, 2] though ultimately rejected by both). Since the formalisation that occurs is crucial to the application of Gödel's first incompleteness theorem to informal mathematics, it would be very convenient for

[8]The 'many' here is due to the fact that it might end up being case that multiple informal proofs are mapped to the same formal proof.

[9]I use the terms 'super-theory' and 'super-system' throughout this paper. I do not intend anything of the 'super-' prefix besides that it is all-encompassing of mathematics in the way described.

Priest's argument if the picture that is sketched here is the correct one, as this would take formalisation to effectively reduce informal mathematics to something formal, and thereby allow the argument to proceed.

Unfortunately, we have good reason to think that this picture cannot be correct. It is obvious that tidying-up syntax is not going to be a many-one mapping. If we start with the natural-language versions of our mathematical theorems, there will be a whole selection of ways in which we can reproduce these theorems in some particular formal language. Even translating very simple fragments of mathematics into simple formal systems can easily lead to a plurality of results. Scaling this up to include *all* of mathematics exacerbates this problem significantly. Add to that the fact that we don't start with a particular formal language that we are to be translating the informal into, but instead generate it "on the fly" based on the syntax of our informal mathematics. That there will only be one possible result is clearly absurd.[10]

Note also that the conversion of informal mathematics into this super-theory is not really like the standard conversion of informal mathematics into some 'foundational' theory such as ZFC set theory (which is potentially what the mathematicians that Priest invokes might have in mind). For if this were the case we would quickly find ourselves with the Benacerrafian problem that there are a large number of different adequate representations for our informal concepts (see [5]). This would lead us out of the first option and its super-theory, into a picture where there are multiple different formalisations of informal mathematics.

I would like to emphasise here that the worry I am raising with the generated super-theory is nothing to do with its inconsistency (for such a theory would undoubtedly be inconsistent) and as such it is not open to the usual charge of begging the question against the dialetheist.

5.2 A Second Option

As a second option, Priest could hold it that the formalisation process for all mathematics that he is after is actually a case of *many-many formalisation*. As I have already argued, there may be many different formalisations of mathematics, which Priest can accept as the case in order to avoid the problems presented against the many-one formalisation picture. In essence, this approach is embracing the plurality of formalisations as opposed to letting it become a problem.

However, accepting this path immediately adds an extra complication to the argument, in that now Priest's claims about the formalised version of informal proof must implicitly be quantifying over formalisations. In particular, each time he mentions

[10] An anonymous referee suggests that we may be able to distinguish between a plurality of results which are equivalent under translation and those which genuinely disagree. I believe, however, that this will not save the argument. In a critical discussion of Azzouni's formalist account of proofs [20], I have previously argued that such a move is not going to deliver the substantial kind of formalisation required for the argument to proceed.

the formalised version of a proof of informal mathematics, there is no one thing this refers to but instead a selection of different formalised versions of the informal proof. The next natural question to follow this up with is how to determine which formalisations fall under this quantification for any given proof. Put another way: which formalisations of informal mathematics will be adequate and acceptable? For example, a formal language which is too expressively weak to even state standard theorems would be inadequate and unacceptable. The question, then, comes down to finding (and defending) criteria of adequacy for these formalisations of informal mathematics.

Formalisation, as it is being conceived of here, is not a process of exposing an underlying logical form already present in the informal proof, or any thought in this direction. I take this to be the case because informal proofs will underdetermine the language, system and structure that such a proof would adhere to and have. It is instead taken to be a process that is inextricably linked to the context in which it occurs. Relevant factors include the agent performing the formalisation, their purposes in doing so and the formal theory they intend to formalise the given informal proof into. It might be useful here to consider an analogy to Carnap's notion of explication (as in [6]) where there is also no definitive fact of the matter as to what the correct explication is for some given concept. Instead the different results are compared and evaluated using pragmatic measures such as usefulness, simplicity, explanatoriness, precision etc.

In a similar way, there could be a whole range of formalisations that can be of varying degrees of usefulness in making some informal piece of mathematical reasoning fully formal. In Priest's formalisation of all of informal mathematics we may find a number of different results which are of varying degrees of usefulness, explanation, accuracy, simplicity etc. Of course, amongst these there may be a number of formalisations that we would want to recognise as inadequate, such as that in the above example of an expressively weak language. We want some way of excluding these examples of 'bad' formalisations of informal mathematics from being implicitly quantified over in Priest's argument. However Priest would want to go about this project, we can see that it adds significant philosophical ground that needs to be supplemented to the argument in question before it goes through.[11]

6 On Mathematical Super-Theories

A new worry that emerges from the consideration of different formalisations concerns the reliance on one (or indeed many) mathematical super-theories. Since we have seen the analogy to Carnap and want to evaluate our formalisations using pragmatic principles, we must consider whether unified mathematical super-theories, in the

[11] An anonymous referee proposes an additional argument against Priest based on this section: that the translation on the many-many case is not effective means that informal proof can therefore not meet the minimum requirements for falling under Gödel's theorems. Grist to the mill!

sense that Priest has proposed, are indeed the best when evaluated in this way. In this subsection I will briefly consider three reasons why this might not be the case.

Before I begin, though, let us just make explicit why for Priest's argument there is now the need to formalise all of informal mathematics in one go, in its entirety, into a super-theory. If this is not done another key step of the argument cannot go through, namely the step where it is insisted that the Gödel sentence is indeed provable. If we were to replay the argument just in arithmetic, for example, we would code in (the formalisation of) *informal provability in arithmetic* and soon discover the Gödel sentence is not provable in this formalisation. But here we would be free to take the traditional lesson that this is just a limitation on the formalisation, which may well be incomplete.[12] It is only be squeezing out all room for this incompleteness by quantifying over all mathematics and informal proof simpliciter that the argument could hope to successfully establish that the answer is actually inconsistency rather than mere incompleteness.[13]

Let us now consider why this super-theory will run into difficulties.

One worry may be that different fields or areas of mathematics might be best served by different formal systems, or even different styles of formal systems. For example, the study of algebra, set theory and geometry all appear very different at first glance, and so it may be that they are best served by being formalised into different formal systems (say, with different proof rules which better track the kinds of inferences made in these fields). Of course, the judgment here must be relative to some purpose of formalisation, but we may take the purpose at hand to be (something like) giving a formal reconstruction of the informal proofs, which tracks the inferential steps that were being used. To justify this, recall that Priest's treatment of informal mathematics as a formal theory was meant to be "without injustice".

The first problem I am proposing, then, is that it might be that different formal systems, that are tailored to different sub-areas of mathematics, might allow the more accurate reconstruction of the reasoning present in the informal proofs for those different areas. It also seems that Priest cannot point to the fact that the super-system(s) he is after are those that represent a "tidying up' of the fragment of natural language that mathematics is expressed in, because the point that is being pressed here is that this talk is an over-simplification of a more complex process.

Relatedly, the second concern I have is that diagrammatic proofs may lead to a significant worry for Priest. In referring to the "fragment of English" that informal mathematics is expressed in, Priest seems to miss a wide selection of mathematics that is communicated pictorially. Pictures can serve to communicate mathematical facts, but can also function as components of informal proofs or proofs in their entirety (see [13, 14]. How is this to be accommodated in the super-systems which are meant to formalise all of informal mathematics? What will the formalisation process

[12]And we are well used to theories being incomplete for more reasons than Gödel theorem. For instance, Peano arithmetic also has examples like Goodstein's theorem and the Paris-Harrington theorem.

[13]Note that this cannot be avoided by insisting that the Gödel sentence must be part of naïve arithmetic without running afoul of the distinction of Sect. 3.

do to diagrammatic proofs? If they are simply to be eliminated, this once again means that informal mathematics is undergoing a drastic change in the formalisation process. Alternatively, there are formal systems for diagrams which may serve to formalise some of the diagrammatic proofs. However, we are now engaged in a project of making the super-systems, which originally sounded straightforwardly close to informal mathematics, encompass much broader pieces of mathematical reasoning. At the very least, this is a non-trivial undertaking which involves constructing a mixed-mode formal system which combines traditional syntactic components with formal diagrammatics. A deeper worry, however, is that we are now able to question whether it will even be possible to capture all of the mathematical reasoning that occurs in informal proofs in formal systems, without doing violence to the source material. I shall return to this line of thought in Sect. 8.

A third problem we encounter for the mathematical super-theory can draw on Priest's own considerations of mathematical pluralism in Priest [16]. Modern mathematical investigation extends to examining which results obtain from adopting different logics to work in. Yet if all the various investigations of different logics are taken to be part of informal mathematics, what happens when we formalise them into the one super-theory? Not only do we face the prospect of systems collapsing into one another, but the more alarming danger of triviality looms. Observe that some of the logics we might want to use will include the principle of explosion, most notably classical logic. As soon as a contradiction arises somewhere in the system (which is exactly what Priest's argument is attempting to force), immediately it follows that the whole super-system is trivialised. This is regardless of whether we think that there is something *philosophically* wrong with classical mathematics, and the principle of explosion in particular, since we are just formalising informal mathematics as we found it. This worry also doesn't rely on logical pluralism, instead just the more uncontroversial fact of logical plurality.[14] In the case of this worry, Priest's argument will still go through but using the fact that a trivial super-system is also inconsistent, which is hardly a desirable result.

7 Fragmented Formalisations

The counter-suggestion to formalising all of informal mathematics simultaneously into one super-theory, with which we have seen some serious difficulties, is that the formalisation process may be one that can only be successful when done in a fragmented way. The suggestion is that constructing a formal system is achievable when we take smaller "chunks" of mathematics that we want to formalise, just not when we want to take it all at the same time. Such an understanding would provide reasonable solutions to dealing with the problems of previous section, without giving up the possibility of formalising parts of mathematical reasoning.

---

[14] I take it that, as mathematicians, we don't need to commit ourselves to the truth, in some philosophical sense, of the mathematics that is being carried out.

Let us see why switching from the idea of a super-theory to the fragmented approach is not a good option if we want to maintain Priest's argument that informal mathematics is inconsistent by Gödel's First Incompleteness Theorem. The issue is that the argument relies on capturing informal mathematics fully to insist that the sentence φ, which is unprovable in the formalised version of informal mathematics but is nonetheless established by informal proof, must also by provable in the formalised system. If, however, it fails to obtain that any one theory does successfully formally represent all of informal mathematics as a whole, then it cannot be insisted that the last step holds. The point is that we get to the fact that the sentence must be true in the system because the system includes all informal mathematical reasoning. If we do not guarantee this, then the inconsistency is not guaranteed either.

Undermining this last step is sufficient for giving a criticism of Priest's argument, but what we have seen so far forms a somewhat deeper difficulty. Priest's more general project in *In Contradiction* is to re-examine the balance between completeness and consistency, insisting that it is the latter we jettison in light of Gödel's theorems rather than the former, which is the orthodox choice. Recall that in Sect. 3 we set aside Beall's use of the same balancing act, where he suggests that when formalising mathematical reasoning we are returned to the completeness/consistency dichotomy. What has implicitly been done here, then, is to use considerations of the process of formalisation to give an independent motivation for why we might prefer to end up with an incomplete system when formalising informal proofs, without making reference to any concerns about consistency.

8 On the Formal and the Informal

For all that has been said, I think there is another more devastating objection to Priest's argument. In Sect. 5.1 we saw that the idea that there would only be one formalised counterpart of informal mathematics would not hold any water. However, it was only on this reading that it seemed acceptable to treat informal mathematics as if it were a formal theory, at least superficially, stemming from the fact that there was one 'body' of informal mathematics and one formalisation thereof. Nonetheless, having been discussing the difficulties involved in formalising theories, it should now be becoming clearer that there was something fishy going on in this step of the argument.

The objection is the following: by moving from informal proof to a formalised version thereof, Priest's argument is guilty of changing the subject. The argument intended to show that informal proof was inconsistent, and not just coincidentally but *inherently* so. Yet, almost immediately in the reasoning, to get the application of the incompleteness results off the ground, Priest needs the subject of his argument to be a formal theory. The answer, therefore, is that mathematics is not a formal theory and that transforming it to be one will do an injustice to its source material. The argument speaks as if the multiple representations that informal mathematics can have as formal systems are identical to the informal mathematics itself, but this is just a confusion of distinct things.

While Priest was looking to demonstrate that informal mathematics was inherently inconsistent, an option that is now on the table is that mathematical reasoning is *inherently informal*, a view common in the mathematical practice literature (e.g. [9]), or that it may be inherently incomplete, or indeed both. The thought would then, in these cases, be that no formal system would suffice to adequately capture mathematics in its entirety. Indeed, this is the traditional lesson that people take from the incompleteness results, but this standard result relies on the question-begging move from consistency to incompleteness. Now, though, we have seen independent motivations for thinking so and rejecting the argument.

Priest's challenge was looking to adjust the balance between consistency and completeness in favour of the latter over the former. But now, by considering the third axis of formality and informality, we have obtained a way to defend incompleteness over inconsistency in the formal setting without begging the question.[15] For the argument relies on a number of assumptions about the nature of formalisation which allow one to easily and without injustice take informal mathematics into formal mathematics. I have, to the contrary, argued that this distinction runs deep and cannot be bypassed lightly, meaning that arguments that work for formal theories cannot be straightforwardly applied to informal mathematics, and ultimately that Priest's argument does not go through.

References

1. Azzouni, J. (2004). The derivation-indicator view of mathematical practice. *Philosophia Mathematica (III)*, *12*, 81–105.
2. Azzouni, J. (2005). How to nominalize formalism. *Philosophia Mathematica (III)*, *13*, 135–159.
3. Beall, J. (1999). From full blooded platonism to really full blooded platonism. *Philosophia Mathematica (III)*, *7*, 322–325.
4. Beall, J. (2009). *Spandrels of Truth*. Oxford: Clarendon Press.
5. Benacerraf, P. (1965). What numbers could not be. *The Philosophical Review*, *74*, 47–73.
6. Carnap, R. (1945). The two concepts of probability. *Philosophy and Phenomenological Research*, *5*, 513–532.
7. Dutilh Novaes, C. (2011). The different ways in which logic is (said to be) formal. *History and Philosophy of Logic*, *32*, 303–332.
8. Hersh, R. (1997). Prove–once more and again. *Philosophia Mathematica (III)*, *5*, 153–165.
9. Larvor, B. (2012). How to think about informal proofs. *Synthese*, *187*, 715–730.
10. Leitgeb, H. (2009). "On Formal and Informal Provability". In O. Bueno, Ø. Linnebo(Eds.), (2009): *New Waves in Philosophy of Mathematics*, Palgrave Macmillan, pp. 263–299.
11. Marfori, M. A. (2010). Informal proofs and mathematical rigour. *Studia Logica*, *96*, 261–272.
12. Milne, P. (2007). On Gödel sentences and what they say. *Philosophia Mathematica (III)*, *15*, 193–226.

---

[15] Or, at least, to see incomplete and inconsistent systems as both serving purposes which may be justified by key pragmatic principles.

13. Nelsen, R. B. (1993). *Proofs Without Words: Exercises in Visual Thinking*. Washington: The Mathematical Association of America.
14. Nelsen, R. B. (2000). *Proofs Without Words II: More Exercises in Visual Thinking*. Washington: The Mathematical Association of America.
15. Priest, G. (1987). *In Contradiction: A Study of the Transconsistent*. Oxford: Oxford University Press.
16. Priest, G. (2012). "Mathematical Pluralism", *Logic Journal of the IGPL*, pp. 4–13.
17. Rav, Y. (1999). Why do we prove theorems? *Philosophia Mathematica (III)*, *7*, 5–41.
18. Rav, Y. (2007). "A critique of a formalist-mechanist version of the justification of arguments in mathematicians' proof practices", *Philosophia Mathematica (III)* 15, pp. 291-320.
19. Robinson, J.A. (1991). "Formal and Informal Proofs" In R.S. Boyer (Ed.),: *Automated Reasoning*, pp. 267–282. Berlin: Kluwer Academic Publishers.
20. Tanswell, F.S. (2015). "A Problem with the Dependence of Informal Proofs on Formal Proofs", *Philosophia Mathematica (III)*, Vol. 23(3), currently available through advance access.

On the Methodology of Paraconsistent Logic

Heinrich Wansing and Sergei P. Odintsov

Abstract The present note contains a critical discussion of the methodology of paraconsistent logic in general and "the central optimisation problem of paraconsistent logics" in particular. It is argued that there exist several reasons not to consider classical logic as the reference logic for developing systems of paraconsistent logic, and it is suggested to weaken a certain maximality condition that may be seen as essential for "optimisation", which is a methodology in the tradition of Newton da Costa. It is argued that the guiding motivation for the development of paraconsistent logics should be neither epistemological nor ontological, but informational. Moreover, it is pointed out that there are other notions of maximality and other methodologies. A methodology due to Graham Priest and Richard Routley and another methodology that focuses on a minimal shrinkage of expressiveness relative to a given reference logic are considered in some detail.

Keywords Paraconsistent logic · Methodology · Maximal paraconsistency · Classical logic · Constructive logic · Connexive logic · Absorption · Relevance logic · Separation of concepts · Minimal loss of expressiveness

H. Wansing (✉)
Ruhr University Bochum, universitätsstr 150, 44801 Bochum, Germany
e-mail: Henrich.Wansing@rub.de

S.P. Odintsov
Sobolev Institute of Mathematics, Novosibirsk State University, Novosibirsk, Russia

H. Andreas and P. Verdée (eds.), *Logical Studies of Paraconsistent Reasoning in Science and Mathematics*, Trends in Logic 45,
DOI 10.1007/978-3-319-40220-8_12

1 Introduction

The description of the scientific goals of the recent conference on *Paraconsistent Reasoning in Science and Mathematics*[1] circulated to the invited speakers of the conference, contains the following considerations on the methodology of paraconsistent logic:

> The variety of approaches to paraconsistency raises two obvious questions: what are the desiderata that a paraconsistent logic ought to satisfy? Which paraconsistent logics score well given certain desiderata? Regarding the first question, there is some consensus on there being three core desiderata:
>
> (1) A paraconsistent logic ought to capture the inferential use of inconsistent but non-trivial theories.
>
> (2) A paraconsistent approach should explain how one can weaken the underlying logic of classical logic to get rid of the explosion principle and still have enough inferential power to be successful.
>
> (3) It is desirable to have a philosophical motivation for the deviation from classical logic in terms of epistemological and, possibly, also metaphysical considerations.
>
> It is obvious that full preservation of classical logic is incompatible with satisfaction of the first desideratum. So, we are confronted with an optimisation problem. It is no exaggeration to call it the central optimisation problem of paraconsistent logics. This problem, of course, may admit of more than one solution. Also, it is not clear whether a single and universal solution is achievable or even desirable.

This paper contains a discussion of some methodological aspects of paraconsistent logic, and a special emphasis is put on "the central optimisation problem of paraconsistent logics". It is argued that there are several reasons not to consider classical logic as the reference logic for developing systems of paraconsistent logic, and it is suggested to weaken a certain maximality condition that may be seen as essential for "optimisation". Moreover, irrespective of how strongly paraconsistent reasoning may be motivated by applications to knowledge representation or by ontological considerations, it is argued that for logic as a theory of valid inference, the guiding motivation for the development of systems of paraconsistent logic should be neither epistemological nor ontological, but *informational*. We also briefly discuss the idea of *ex contradictione nihil sequitur* as another conception of maximality, consider a methodology of paraconsistent logic due to Priest and Routley [57], and present a methodology that imposes a condition of minimal loss of expressiveness relative to a given reference logic.

---

[1]See http://www.paraconsistency2014.philosophie.uni-muenchen.de/index.html, Munich, June 11–13, 2014. Although this description was not intended for publication, we include it here because it presents methodological views many paraconsistent logicians seem to agree with and which therefore may serve as a suitable starting point for our methodological considerations.

2 Constraints and Desiderata: A Brief History

The methodological discussion of constraints on and desiderata to be fulfilled by systems of paraconsistent logic has some tradition, and presenting it partly explains why the methodological views quoted in the introduction seem to be fairly representative. The above desiderata (1)–(3) to some extent echo the list of conditions stated by Stanisław Jaśkowski in his seminal paper on paraconsistent logic from 1948:

> Accordingly, the problem of the logic of inconsistent systems is formulated here in the following manner: the task is to find a system of the sentential calculus which: ⟨1⟩ when applied to the inconsistent systems would not always entail their overfilling, ⟨2⟩ would be rich enough to enable practical inference, ⟨3⟩ would have an intuitive justification. ([33, p. 38])

Jaśkowski admits that it is difficult to assess the third condition objectively.[2]

Newton Da Costa [19, p. 498] lists a number of conditions his propositional calculi C_n, $1 \leq n \leq \omega$, ought to satisfy if they are to serve as logical bases for non-trivial inconsistent theories, where C_0 is classical propositional logic:

> (I) In these calculi the principle of contradiction, $\neg(A\&\neg A)$, must not be a valid schema; (II) From two contradictory formulas, A and $\neg A$, it will not in general be possible to deduce an arbitrary formula B; (III) It must be simple to extend C_n, $1 \leq n \leq \omega$, to corresponding predicate calculi (with or without equality) of first order; IV) C_n, $1 \leq n \leq \omega$, must contain the most part of the schemata and rules of C_0, which do not interfere with the first conditions. (Evidently, the last two conditions are vague.)

Da Costa and Alves [20, p. 185] present the following conditions:

> (1) From two contradictory formulas, P and $\neg P$, it should not be possible in general to deduce an arbitrary formula.
> (2) The system should contain most of the schemata and deduction rules of the classical calculus that do not inference[3] with the first condition.

The conditions (1), ⟨1⟩, (I), (II), and (1) may be seen as unproblematic because they define the subject, although nowadays (I) is usually not used in definitions of paraconsistency. There is room for being critical about assuming the availability of conjunction in (I) and room for discussing (I), (II), and (1) by asking which characterization of contradictoriness is assumed or whether $\neg$ may also be an operation that gives rise to contrary instead of contradictory pairs of formulas, cf. [67], but

---

[2]Alexander Karpenko [36] suggests a concrete reading of conditions ⟨2⟩ and ⟨3⟩. He requires that a propositional paraconsistent logic should validate *modus ponens* and the three schemata characteristic of BCI-logic:

(I) $p \to p$,
(B) $(q \to r) \to ((p \to q) \to (p \to r))$,
(C) $(p \to (q \to r)) \to (q \to (p \to r))$.

Moreover, in a three-valued semantics, the truth tables for negation, implication, conjunction, and disjunction should coincide with the classical tables on the classical values 0 and 1.

[3]This should probably be “interfere”.

the general idea is uncontentious. It is assumed that the language under consideration contains a *negation* connective, and a theory containing both a formula and its negation should not, in general, be trivial. Condition (2) is much less uncontroversial; it declares classical logic as the reference logic. The notion of a reference logic has two aspects. One aspect is that a paraconsistent logic ought to be faithful to the reference logic in the sense of being a *subsystem* of it. A second aspect is that a paraconsistent logic should be a *maximally paraconsistent* subsystem of the reference logic. The requirement of maximal paraconsistency can, of course, be detached from considering any particular background logic.

Ofer Arieli, Arnon Avron, and Anna Zamansky [7] take the multitude of different kinds of paraconsistent logics that can be found in the literature as a reason to look for "ideal" paraconsistent propositional logics. Their starting point is the methodology of da Costa, so that an ideal paraconsistent logic preserves as much of classical logic as possible, and they wonder what exactly that constraint amounts to. According to them [7, p. 32] a preliminary analysis reveals three fundamental and intuitive properties:

> **Containment in Classical Logic**. As the general characterization given above to 'ideal paraconsistent logics' suggests, classical logic is usually taken as the reference logic for such logics. This means that while a reasonable paraconsistent logic is necessarily more tolerant than classical logic (since it allows non-trivial contradictions), it should not validate any inference which classical logic forbids. In other words: it should be contained in classical logic.
> **Maximal Paraconsistency**. The requirement from a paraconsistent logic **L** to retain as much of classical logic as possible, while still allowing non-trivial inconsistent theories has two different interpretations, corresponding to the two aspects of this demand:
>
> > **Absolute maximal paraconsistency**. Intuitively, this means that by trying to further extend **L** (without changing the language) we lose the property of paraconsistency.
> > **Maximality relative to classical logic**. Here the intuitive meaning is that **L** is so close to classical logic, that any attempt to further extend it should necessarily end up with classical logic.
>
> Ideally, we would like of course an ideal paraconsistent logic to have both types of maximality.[4]

[4]That *the combination* of absolute maximal paraconsistency (or some other notion of maximal paraconsistency not defined with respect to a given reference logic) and maximality relative to classical logic is indeed ideal is not so clear. Even Arieli, Avron, and Zamansky [6, p. 32] refer to maximality simpliciter as the "more natural" notion of maximal paraconsistency as compared to maximality with respect to classical logic. In Chap. 2 of a book manuscript on the theory of effective propositional paraconsistent logics, they change their terminology. The logics they previously referred to as ideal paraconsistent logics are now called "fully maximal and normal". The reason for that choice of a more neutral term, as they told us, is that on the one hand in their book they consider approaches to paraconsistency different from da Costa's and on the other hand they regard the class of all "ideal" paraconsistent logics as still too broad to indeed justify designating all these logics as ideal.

Reasonable language. Obviously, the language of a paraconsistent logic should have an official negation connective which is entitled to this name. This is insufficient, of course. Thus, in [3] we have shown that the three-valued logic whose only connective is Sette's negation [32], is maximally paraconsistent and it is obviously contained in classical logic. Still, nobody would take it as an ideal paraconsistent logic, because its language is not sufficiently expressive. So an ideal paraconsistent logic should be in a language which is reasonably strong.

In their [5, p. 706], Arieli, Avron, and Zamansky present four properties as "desirable for a decent logic for reasoning with inconsistency," the first is paraconsistency and the second is more explicit about a reasonably strong language:

1. Paraconsistency. The rejection of the principle of explosion, according to which any proposition can be inferred from an inconsistent set of assumptions, is a primary condition for properly handling contradictory data.
2. Sufficient expressive power. Clearly, a logical system is useless unless it can express non-trivial, meaningful assertions. In our framework, a corresponding language should contain at least a negation connective, which is needed for defining paraconsistency, and an implication connective, admitting the deduction theorem.
3. Faithfulness to classical logic. As observed by Newton da Costa, one of the founders of paraconsistent reasoning, a useful paraconsistent logic should be faithful to classical logic as much as possible. This implies, in particular, that entailments of a paraconsistent logic should also be valid in classical logic.
4. Maximality. The aspiration to retain as much of classical logic as possible, while still allowing non-trivial inconsistent theories is reflected by the property of maximal paraconsistency, according to which any extension of the underlying consequence relation yields a logic that is not paraconsistent anymore.

They refer to logics that satisfy the four properties as *ideal* for reasoning with inconsistency and develop a precise definition of ideal paraconsistency. In Sect. 5 we will take up their general setting for discussing the methodology of paraconsistent logic, their condition of expressive strength, and the rather general condition they impose on negation connectives. As already indicated, we will, however, criticize their choice of classical logic as a reference logic and their notion of maximal paraconsistency.[5]

3 Motivation in Terms of Epistemological or Metaphysical Considerations

Let us first consider the above listed desideratum (3) and only later consider the question of classical logic as a reference logic, the notion of maximality, the idea of *ex contradictione nihil sequitur*, and methodologies different from da Costa's. We believe that logic should avoid as many ontological commitments as possible.

[5]The depth and beauty of their results is, however, undoubted.

With the conception of logic as the theory of valid inferences and the conception of logics as consequence relations, logic, by definition, is committed to the existence of languages but not necessarily to the existence of language users. Furthermore, if models are to be constructed, then they are built from the unavoidable linguistic entities. Moreover, the notion of valid inference does not refer to the knowledge or belief states of any epistemic or doxastic subjects. That is not to say that logic does not have an epistemic dimension as well. We "draw" conclusions and in doing so form beliefs and acquire knowledge. The exact nature of this process is complicated,[6] and it is not at all obvious that systems of paraconsistent logic should be motivated in terms of epistemological considerations or metaphysical ones.

Also, the motivation of a system of paraconsistent logic may well lead to abandoning the characterization of entailment in terms of truth preservation. Nevertheless, if a convincing motivation for a system of paraconsistent logic can be given without appeal to language users, epistemic subjects possessing mental states, or without appeal to dialetheia, i.e., true contradictions, then *prima facie* such a motivation is to be preferred over a motivation that comes with the mentioned epistemological or ontological commitments. That does not mean that a motivation in epistemological or ontological terms or by applications to knowledge representation may not be very useful, but nevertheless it may be seen as an advantage if a paraconsistent logic can be motivated independently of epistemological or ontological commitments.

A motivation that omits such commitments can be given in terms of information. There exist various notions and theories of information, but for our present purposes it suffices to share Michael Dunn's [24, p. 423] general and basic view of information: "information is what is left of knowledge when one takes away belief, justification, and truth. ... Information is ... a kind of semantic content – the kind of thing that can be expressed by language." In [25, p. 582], Dunn explains that information is something like a Fregean thought. Information so understood clearly is not veridical and thus has to be distinguished from semantic information as understood by, for example, Luciano Floridi, see the surveys [2, 28] and the references given there. Having said that much, we may suppose that collections of assumptions (sets, multisets, lists, or even more complex structures of formulas) provide information even if this information happens not to be processed by any human or artificial information processor. Let us, for determinacy, consider sets. A consequence relation holds between a set and another set of formulas, an at most single-element or possibly multi-element set. If a formula A (or rather the singleton $\{A\}$) is entailed by or derivable from the set of premises Δ, then Δ provides the information that A, irrespective of whether that is recognized or not by some epistemic subject. If we assume a language containing a *negation* operation $\sim$, then we assume that there exists some kind of semantic opposition between A and $\sim A$. What this opposition amounts to and how exactly it is to be spelled out in the semantics of a paraconsistent logic is a fundamental question. In any case the very idea of paraconsistent logic, namely that the schematic inference from $\{A, \sim A\}$ to an arbitrary formula B (or, more generally, the schematic inference from $\Delta \cup \{A, \sim A\}$ to Γ for arbitrary sets

[6] See [51].

of formulas Δ, Γ) is invalid, receives a very intuitive motivation if it is phrased in terms of information: in general, it just is not the case that $\{A, {\sim}A\}$ provides the information that B ($\Delta \cup \{A, {\sim}A\}$ provides the information that Γ), for arbitrary B (Δ and Γ). If we employ a conjunctive reading of premise sets and a disjunctive reading of conclusion sets, then in perfect duality to the case of inconsistent premise sets we obtain a very intuitive motivation for the invalidity of the entailment from Δ to $\Gamma \cup \{A, {\sim}A\}$ for arbitrary sets of formulas Δ, Γ: in general, it just is not the case that Δ provides the information that $\Gamma \cup \{A, {\sim}A\}$).[7]

The intuitive, pre-theoretical understanding of information seems to admit both partial and conflicting "pieces" of information. A piece of information is partial if for at least one proposition (expressed by the formula) A it neither provides the information that A nor the information that ${\sim}A$. A piece of information provides conflicting information if for at least one proposition A it provides not only the information that A but also the information that ${\sim}A$.

Not only strike us these motivating ideas as simple and clear, they are also apt to logic as a discipline free from unnecessary epistemological (and ontological) commitments. As Jon Barwise [12, p. 368] once put it, "[i]nformation travels at the speed of logic, genuine knowledge only travels at the speed of cognition and inference."

4 Classical Logic as the Reference Logic for Paraconsistent Logics

Before turning to maximal paraconsistency, we shall next discuss faithfulness to classical logic and maximality relative to it, i.e., the notion of a reference logic and of faithfulness presented in [5, p. 706]. In the mentioned declarations of classical logic as the reference logic for the development of paraconsistent logics, the choice of classical logic is more or less taken for granted, although one may wonder why exactly a *nonclassical* paraconsistent logic, if correct, should have a distinguished status in virtue of being faithful to *classical* logic "as much as possible".

Classical logic is a natural logic for reasoning about what is and what is not the case in classical, platonistic mathematics. It is not, however, a natural reference logic (in the mentioned sense) for reasoning about the universe of constructive mathematical objects. Moreover, classical logic is not at all a natural reference logic for reasoning about information and information structures. On the other hand, it is reasoning about information that suggests paraconsistent reasoning.

We will present four considerations that together may cast doubt on using classical logic as the reference logic for developing systems of paraconsistent logic: (i) Clearly, a reference logic is a distinguished logic, and being classical is a distinction. How-

[7] The guiding ideas of relevance logic and containment logic are in the same spirit but far more specific, see, for example, [4, 27, 42, 65].

ever, the classicality of classical logic may be seen as a contingent historical fact.[8] (ii) The fundamental motivation of classical logic is in conflict with the motivation of paraconsistent logic as a logic of information structures. (iii) Classical, Boolean implication tends to be in conflict with paraconsistent reasoning. (iv) It is not clear that closeness to classical logic is justified by the intended applications of paraconsistent logic. In view of these observations, it is doubtful that the choice of classical logic as a reference logic is indeed justified.

Ad (i). Indeed, many important non-classical logics in the vocabulary of classical logic are subsystems of classical logic. Avron [11, p. 1] remarks that "in practice almost all non-classical logics ever seriously studied" are subsystems of classical logic. There is a continuum of superintuitionistic logics that are intermediate between intuitionistic and classical logic, and, as an anonymous referee highlighted, any non-trivial supersystem of intuitionistic logic in its vocabulary is a subsystem of classical logic. Systems of relevance logic and other substructural logics are subsystems of classical logic, and in [11] it is shown that language-preserving extensions of almost every relevance logic in the tradition of Anderson and Belnap [4] can have only classical tautologies as theorems. Classical logic seems to be the upper limit for logical systems in its language. The fact that there exist "contra-classical" logics the connectives of which cannot be (definitionally) translated so as to obtain a subsystem of classical logic (see [32]) might, perhaps, be seen as a fact that only slightly disturbs the general picture. But maybe the threatening disturbance is not so small, also from a historical perspective. It seems that Aristotle and Boethius advocated an understanding of negated implications that is orthogonal to classical logic. In so-called systems of connexive logic, certain non-theorems of classical logic are taken to be valid, in particular, "Aristotle's Theses"

$$\sim(\sim A \to A) \text{ and } \sim(A \to \sim A).$$

and "Boethius' Theses"

$$(A \to B) \to \sim(A \to \sim B) \text{ and } (A \to \sim B) \to \sim(A \to B).$$

The validity of these schemata involving negated implications can be motivated in various ways,[9] and the classical understanding of negated implications expressed by the bi-conditional $\sim(A \to B) \leftrightarrow (A \wedge \sim B)$ may be seen critically. If one, for instance, assumes the (contentious) denial equivalence view, according to which the denial of a statement A can be adequately analyzed as the assertion of A's negation,[10] then it is not obvious that denying $(A \to B)$ is to be analyzed as a compound speech act consisting of the assertion of A and the denial of B.

[8] We do not intend to deny that platonistic mathematics and the entire realistic tradition in ontology, metaphysics, and science play an important role in the history of ideas.

[9] Cf. [43, 69, 70].

[10] See [60].

Moreover, also in terms of its meta-theory, the classicality of classical predicate logic might turn out to be accidental. We are so used to the fact that classical predicate logic is undecidable that we are not much inclined to see this property as being in conflict with classical predicate logic's distinction as classical. There are, however, decidable subsystems of classical predicate logic in its full vocabulary, see [37] and [38]. Classical (and intuitionistic) logic have presentations as sequent calculi such that dropping the contraction rule results in decidable systems. Moreover, giving up contraction has the pleasant side-effect of getting rid of the Curry paradox. If logic had not, as a matter of fact, been developed for classical mathematics but for modelling resource-sensitive reasoning, the contraction-free, affine subsystem of classical first-order logic might well have emerged as the logical orthodoxy.

Paraconsistency does deviate from logical orthodoxy, but it is not at all clear that classical logic indeed is the logical orthodoxy from which paraconsistent logics ought to deviate only minimally. The first-order extension **QN4** of Nelson's *constructive* paraconsistent logic **N4** with strong negation, for example, is a subsystem of classical predicate logic and is faithfully embeddable into positive first-order intuitionistic logic. **QN4** is constructive insofar as it enjoys the disjunction property, the constructible falsity property and the existence property. Why should one require that **QN4** deviates only minimally from classical first-order logic **QCL**? It is even less plausible to require that the paraconsistent *connexive* first-order logic **QC** from [69] deviates only minimally from **QCL**. Arieli, Avron, and Zamansky's requirement of being "faithful to classical logic as much as possible" is not convincingly justified.

Ad (ii). Classical logic is a logic for realists; it is a logic for reasoning about what is and what is not the case.[11] Classical negation is the Boolean complement, and the classical negation $\neg A$ of A treats falsity as the absence of truth. If we deal with information about what is or is not the case, falsity and the absence of truth fall apart and we are in an at least four-valued setting. As already remarked, information with respect to a given proposition may be partial or conflicting; it may even be both, cf. [62].

Ad (iii). The conflict between disjunction introduction, disjunctive syllogism, and paraconsistent reasoning has been observed several times by various authors and can be traced back to the logician(s) from the 14th century called "Pseudo-Scotus".[12] It is very nicely described, for example, in Zach Weber's entry on paraconsistent logic for the *Internet Encyclopedia of Philosophy*. Weber explains:

---

[11]There are several, sometimes subtly nuanced notions of realism. Roughly speaking realism is usually understood as the doctrine that there are entities (of a certain kind) that exists independently of any conscious beings. Dummett [23] characterizes realism semantically by the assumption of bivalence, according to which every meaningful declarative sentence from a certain discourse is either true or false but not both true and false and not neither true nor false, provided it is neither vague nor ambiguous. In that sense realism is tied to classical logic, and it would be incorrect, for example, to classify dialetheists as realists.

[12]In [18, Footnote 3] John of Cornwall is mentioned as the most probable author.

> But we cannot simply remove the inference of explosion from classical logic and automatically get a paraconsistent logic. The reason for this, and the main, serious constraint on a paraconsistent logic, was discovered by C.I. Lewis in the 1950s. Suppose we have both A and $\neg A$ as premises. If we have A, then we have that either A or B, since a disjunction only requires that one of its disjuncts holds. But then, given $\neg A$, it seems that we have B, since if either A or B, but not A, then B. Therefore, from A and $\neg A$, we have deduced B. The problem is that B is completely arbitrary—an absurdity. So if it is invalid to infer everything from a contradiction, then this rule, called disjunctive syllogism,
>
> $$A \vee B, \neg A \vdash B$$
>
> must be invalid, too.

If we consider the following variant of the *modus tollendo ponens*

$$\neg A \vee B, A \vdash B$$

it is clear that disjunction introduction and *modus ponens* for Boolean implication $\neg A \vee B$ result in trivializing inconsistent premise sets. If we consider disjunction introduction as unrelated to the problem of reasoning from inconsistent assumptions, then, because the deduction theorem (in both directions) and cut ensure *modus ponens*, the Pseudo-Scotus or Lewis argument shows that for a paraconsistent transitive consequence relation and an implication satisfying the deduction theorem, implication cannot be Boolean implication.[13]

Ad (iv) Paraconsistent logics are logics that can be fruitfully applied to reasoning with inconsistent data, and it is not really surprising that paraconsistent, inconsistency-tolerant reasoning has become increasingly important in knowledge representation and in artificial intelligence research in general.[14] The relevance of paraconsistent logics for modelling everyday reasoning has been realized already in the 1990 s when, for example, systems of paraconsistent circumscription and minimally inconsistent **LP**, **LP**m, a non-monotonic version of Graham Priest's Logic of Paradox, have been suggested, see [21, 40, 53]. If reasoning is granted to be both paraconsistent and defeasible, and if in the intended application areas classical logic fails, it is more than doubtful that *classical logic* is justified as a reference logic for the development of paraconsistent logics, even if in discussions of **LP**m the idea of "classical recapture" is sometimes seen as indicating the use of classical logic as a reference logic. One notion of recapture compatible with [53] could be defined as follows. Let $\mathbf{L}_1$ and $\mathbf{L}_2$ be two distinct logics with the same language $\mathcal{L}$ and with consequence relation $\vdash_1$ and $\vdash_2$, respectively. Then $\mathbf{L}_1$ recaptures $\mathbf{L}_2$ iff there exists a family of sets of $\mathcal{L}$-formulas such that for every element Δ from that family, $\{A \mid \Delta \vdash_1 A\} = \{A \mid \Delta \vdash_2 A\}$. If a logic $\mathbf{L}$ recaptures classical logic, $\mathbf{L}$ enjoys classical recapture. If, for example, there is a classically inconsistent set of formulas Δ which is also $\mathbf{L}$-trivial, then

[13] The invalidity of *modus ponens* (alias detachment) in the form of *modus tollendo ponens* has recently been argued for by Jc Beall [15] based on a distinction between logic and rational theory change.

[14] We have contributed to this area with [48, 49].

L recaptures classical logic via the family $\{\Delta\}$. So one may want to impose some conditions on the family of sets with respect to which recapture obtains. There is no space here to discuss notions of recapture and the various possible attitudes towards classical recapture; a very illuminating and careful discussion can be found in [1]. In our opinion, the point of classical recapture is reconstructing the classically valid reasoning that is acceptable according to the ideas that motivate the logic by which classical logic is recaptured. Classical logic is not the canon and yardstick here, but rather it is the non-classical conception of inference that helps identify an acceptable fragment of another consequence relation. The consequence relations of **LP***m* and classical logic, for example, coincide on classically consistent premise sets [53], so that a certain fragment of classical logic is identified as acceptable from the point of view of **LP***m*.

In inconsistency adaptive logics (see, for example, [14, 63] and the references given there), the so-called upper limit logic sets the standards of logical normality. In standard format an adaptive logic is given by a lower limit logic that remains stable in the reasoning process, a set of abnormalities, and an adaptive strategy for minimizing abnormalities. The set of abnormalities is chosen so that adding the axioms that trivialize abnormal theories to the lower limit logic gives the upper limit logic. In other words, we chose abnormalities among formulas that imply everything in the upper limit logic. Note that this approach does not come with a binding commitment to classical logic as the upper limit logic. As a matter of historical fact, the upper limit logic and the set of abnormalities of the most prominent adaptive logics $\mathbf{CLuN}^r$ and $\mathbf{CLuN}^m$ are classical logic together with the set of formulas of the form $A \wedge \neg A$.

One might object (and an anonymous referee did so) that for certain applications, inferences such as disjunctive syllogism are needed that are typically valid in classical logic but invalid in paraconsistent logics. The case of disjunctive syllogism at least does not dictate minimal divergence from classical logic. In the paraconsistent logic $\mathbf{N4}^{\perp}$ [46, 47] the defined intuitionistic negation satisfies disjunctive syllogism, whereas the primitive strong negation does not.

There are at least two questions that remain to be addressed:

1. If classical logic is not justified as a reference logic for defining systems of paraconsistent logic, should there at all be such a reference logic?
2. Is the standard notion of maximality unproblematic even if it is detached from classical logic or any other suggested reference logic?

A positive answer to the first question needs justification and gives rise to a plethora of other questions. Should the choice of a reference logic depend on the possession of other properties such as decidability, the disjunction property, the constructible falsity property, the existence property, the existence of a cut-free sequent calculus (of a certain kind), ...? Should the choice of a reference logic depend on particular applications? We shall not address these questions here but instead first turn to the notion of maximality. We will then consider two different methodologies, one due to Priest and Routley and another one focusing on a minimal loss of expressiveness instead of maximal faithfulness with respect to some given reference logic.

5 Maximality

Arieli, Avron, and Zamansky's notion of maximality is strong insofar as it is defined with respect to *the entire language* of the paraconsistent logic in question. In particular, on pain of violating paraconsistency, there is no room for adding theorems or sequents that are *negation-free*. Since paraconsistency is defined in terms of a unary negation connective, $\sim$, and derivability (or, semantically, entailment), it seems justified to consider in the first place maximality with respect to formulas containing at most occurrences of $\sim$, i.e., to restrict attention to the formulas of the language $\mathcal{L}^{\sim}$, based on a fixed denumerable set At of atomic formulas (alias sentence letters). If, in addition, the natural requirement is imposed that the language contains an implication, $\rightarrow$, satisfying the deduction theorem, it seems justified to consider sets of formulas in the language $\mathcal{L}^{\sim,\rightarrow}$ based on At.

In this section, we first present and adopt the general inferential framework of [5, 7]. We then generalize the notion of maximal paraconsistency and present some increasingly complex examples of maximally paraconsistent logics. We close the section with a tentative definition of desirable paraconsistent logics.

Let $\mathcal{L}$ be a propositional language based on At, and let $Fm(\mathcal{L})$ be the set of all $\mathcal{L}$-formulas. We use $p, q, r, p_1, p_2, \ldots$ as schematic sentence letters, $A, B, C, A_1, A_2, \ldots$ to denote $\mathcal{L}$-formulas, and $\Delta, \Gamma, \Delta_1, \Delta_2, \ldots$ to denote subsets of $Fm(\mathcal{L})$.[15] We write Δ, A (Δ, Γ) instead of $\Delta \cup \{A\}$ $(\Delta \cup \Gamma)$.

Definition 1 A *Tarskian consequence relation* for a language $\mathcal{L}$ is a binary relation $\vdash$ between sets of $\mathcal{L}$-formulas and single $\mathcal{L}$-formulas satisfying:

Reflexivity: if $A \in \Delta$, then $\Delta \vdash A$
Monotonicity: if $\Delta \vdash A$ and $\Delta \subseteq \Gamma$, then $\Gamma \vdash A$
Transitivity (Cut): if $\Delta \vdash A$ and $\Gamma, A \vdash B$, then $\Gamma, \Delta \vdash B$

Definition 2 A Tarskian consequence relation $\vdash$ for a language $\mathcal{L}$ is *structural* iff for every uniform $\mathcal{L}$-substitution θ, $\Delta \vdash A$ implies $\theta(\Delta) \vdash \theta(A)$. The relation $\vdash$ is said to be *non-trivial* iff there exists a set $\Delta \subseteq Fm(\mathcal{L})$ and $A \in Fm(\mathcal{L})$ with $\Delta \nvdash A$; $\vdash$ is called *trivial* iff it is not non-trivial. The relation $\vdash$ is called *finitary* iff $\Delta \vdash A$ implies that there exists a finite $\Gamma \subseteq \Delta$ with $\Gamma \vdash A$.

Definition 3 A *propositional logic* is a pair $(\mathcal{L}, \vdash)$ such that $\mathcal{L}$ is a propositional language and $\vdash$ is a structural, non-trivial, and finitary consequence relation for $\mathcal{L}$.

Definition 4 Let $\mathbf{L} = (\mathcal{L}, \vdash)$ be a propositional logic and let $\mathcal{L}$ contain the unary connective $\sim$. The logic $\mathbf{L}$ is called $\sim$-para*consistent* iff there are formulas A, $B \in Fm(\mathcal{L})$ with $A, \sim A \nvdash B$.[16] $\mathbf{L}$ satisfies *double-negation introduction* (dni) iff $A \vdash \sim\sim A$ and it satisfies *double negation elimination* (dne) iff $\sim\sim A \vdash A$.

---

[15]We will not pay much attention to the mention-use distinction when there is no risk of creating misunderstandings.

[16]Equivalently, one may restrict this condition to sentence letters or require that for every sentence letters p, q it holds that $p, \sim p \nvdash q$. The requirement that there are formulas A, $B \in Fm(\mathcal{L})$ with

It is quite standard to characterize implication in terms of the deduction theorem, i.e., in terms of the familiar right introduction rule for implication in the sequent calculus and its converse. Arieli et al. [6, 7] impose a not so standard condition on negation operators, that differs from the constraints on Avron's "perfect negations" [9, 10] but can be found, for example, in [41]; one direction of the condition is suggested in [39].[17] They refer to the so defined operations as weak negations; for uniformity of terminology we will call them "proper negations".

Definition 5 Let $\mathbf{L} = (\mathcal{L}, \vdash)$ be a propositional logic. A unary connective $\sim$ from $\mathcal{L}$ is a *proper negation* for $\mathbf{L}$ iff there exist formulas A and $B \in Fm(\mathcal{L})$ with: $A \nvdash \sim A$ and $\sim B \nvdash B$. A binary connective $\rightarrow$ from $\mathcal{L}$ is a *proper implication* for $\mathbf{L}$ iff for every $A, B \in Fm(\mathcal{L})$ and every $\Delta \subseteq Fm(\mathcal{L})$ the following holds: $\Delta, A \vdash B$ iff $\Delta \vdash (A \rightarrow B)$.

Definition 6 ([5]) Let $\mathbf{L} = (\mathcal{L}, \vdash)$ be a $\sim$-paraconsistent propositional logic. The logic $\mathbf{L}$ is *maximally paraconsistent in the weak sense* iff every logic $(\mathcal{L}, \Vdash)$ that extends $\mathbf{L}$ without changing the language (i.e., $\vdash \subseteq \Vdash$), and whose set of theorems properly includes that of $\mathbf{L}$, is not $\sim$-paraconsistent. The logic $\mathbf{L}$ is *maximally paraconsistent* iff every logic $(\mathcal{L}, \Vdash)$ that properly extends $\mathbf{L}$ without changing the language (i.e., $\vdash \subset \Vdash$), is not $\sim$-paraconsistent.[18]

Arieli et al. [5, 7] define other notions of maximal paraconsistency as well, in particular the notion of maximal paraconsistency relative to classical logic and the notion of an ideal paraconsistent logic.

Definition 7 ([5]) Let $\mathbf{L} = (\mathcal{L}, \vdash)$ be a propositional logic and let $\mathcal{L}$ comprise a unary connective $\sim$. A *bivalent $\sim$-interpretation for $\mathcal{L}$* is a function $\mathbf{F}$ that associates a two-valued truth table with each connective of $\mathcal{L}$, such that $\mathbf{F}(\sim)$ is the classical truth table for negation. Let $\mathcal{M}_{\mathbf{F}}$ be the two-valued matrix for $\mathcal{L}$ induced by $\mathbf{F}$, and let $\vdash_{\mathcal{M}_{\mathbf{F}}}$ be the structural, non-trivial, and finitary Tarskian consequence relation induced by $\mathcal{M}_{\mathbf{F}}$.

- If $\mathbf{F}$ is a bivalent $\sim$-interpretation for $\mathcal{L}$, then $\mathbf{L}$ is $\mathbf{F}$-*contained in classical logic* if $A_1, \ldots, A_n \vdash_{\mathbf{L}} A$ implies $A_1, \ldots, A_n \vdash_{\mathcal{M}_{\mathbf{F}}} A$.
- $\mathbf{L}$ is $\sim$-contained in classical logic if it is $\mathbf{F}$-contained in classical logic for some $\mathbf{F}$.

$A, \sim A \nvdash B$ may be seen as too weak. This opinion is often justified by pointing to Johansson's minimal logic [34], in which for arbitrary formulas A and B it holds that $A, \sim A \vdash \sim B$. In the book manuscript referred to in Footnote 4, Arieli, Avron, and Zamansky distinguish between $\sim$-paraconsistency and strong $\sim$-paraconsistency, where the latter requires that there are atomic formulas $p, q \in Fm(\mathcal{L})$ with $p, \sim p \nvdash \sim q$. It could make sense to generalize this condition by requiring that for every n-place connective $\sharp$, there are atomic formulas $p, q_1, \ldots, q_n$ with $p, \sim p \nvdash \sharp(q_1, \ldots, q_n)$.

[17]Moreover, Arieli et al. [5–7] do not consider the dual of implication, co-implication. We will refrain from considering co-implication in the present paper; see, however, [71, 72] and the references therein.

[18]Note that in [6], maximally paraconsistent logics are called strongly maximal and maximally paraconsistent logics in the weak sense are called maximally paraconsistent.

Arieli et al. observe that if a logic **L** is $\sim$-contained in classical logic, then $\sim$ is a proper negation for **L**.[19]

Definition 8 ([5]) Let $\mathcal{L}$ be a language with a unary connective $\sim$, and let **F** be a bivalent $\sim$-interpretation for $\mathcal{L}$. Then $A \in Fm(\mathcal{L})$ is a *classical* **F**-*tautology* iff every two-valued valuation, which for every connective $\diamond$ of $\mathcal{L}$ respects the truth table $\mathbf{F}(\diamond)$, satisfies A. A logic $\mathbf{L} = (\mathcal{L}, \vdash)$ is **F**-*complete* iff its set of theorems includes all classical **F**-tautologies.

Definition 9 ([5]) Let **F** be a bivalent $\sim$-interpretation. A logic $\mathbf{L} = (\mathcal{L}, \vdash)$ is **F**-*maximal relative to classical logic* iff the following conditions hold:

- **L** is **F**-contained in classical logic.
- If A is a classical **F**-tautology not provable in **L**, then the addition of A as a new axiom schema results in an **F**-complete logic.

The logic **L** is *maximal relative to classical logic* iff there exists a $\sim$-interpretation **F** such that **L** is **F**-maximal relative to classical logic.

This definition of maximality relative to classical logic is meant to make the idea of faithfulness to classical logic precise.

Definition 10 ([5]) A $\sim$-paraconsistent logic **L** is *ideal* iff it is $\sim$-contained in classical logic, has a proper implication, is maximal relative to classical logic, and is maximally paraconsistent.

Examples and counterexamples

- [5]: Ideal paraconsistent logics are:
  - Sette's logic P_1 [61] (and fragments of P_1 containing Sette's negation),
  - the three-valued logics PAC [8, 13] and J_3 [50],
  - all the 2^{20} three-valued logics considered in [6], including the 2^{13} logics of formal inconsistency introduced in [17].
- [5]: Priest's **LP** [52] is maximally paraconsistent and maximal relative to classical logic but it fails to be ideal because it lacks a proper implication, as shown in [6].
- Nelson's constructive paraconsistent logic **N4** with strong negation $\sim$ [3, 68] and its extension $\mathbf{N4}^{\perp}$ by a falsity constant $\perp$ [46, 47] fail to be ideal $\sim$-paraconsistent logics because they fail to be maximally $\sim$-paraconsistent.
- Johansson's minimal logic [34] fails to be ideal because it fails to be maximally $\sim$-paraconsistent for the paraconsistent minimal negation $\sim$.

[19] In [6] Arieli et al. write that for a unary connective $\sim$ of a logic **L** to deserve the name "negation", **L** necessarily would have to be $\sim$-contained in classical logic.

- Dual-intuitionistic logic and bi-intuitionistic logic [30, 58, 72] fail to be maximally $\sim$-paraconsistent for the paraconsistent co-negation $\sim$, and so does Priest-da Costa logic [26, 55, 56].

Since in the present paper the idea of classical logic as the reference logic for defining paraconsistent logics is rejected, the focus in what follows is on maximal paraconsistency. We deviate from Arieli, Avron, and Zamansky and define a weaker notion of maximal paraconsistency that relativizes maximality to extensions of logics $(\mathcal{L}, \vdash)$ by sequents or formulas from a given subset of $Fm(\mathcal{L})$, for example from $Fm(\mathcal{L}^{\sim})$.

Definition 11 Let $\mathbf{L} = (\mathcal{L}, \vdash)$ be a $\sim$-paraconsistent propositional logic and let $\Delta \subseteq Fm(\mathcal{L})$. A propositional logic $(\mathcal{L}, \Vdash)$ *properly extends* $\mathbf{L}$ *in* Δ iff $(\vdash\restriction\Delta) \subset (\Vdash\restriction\Delta)$. The logic $\mathbf{L}$ is *maximally* $\sim$*-paraconsistent with respect to* Δ iff every propositional logic $(\mathcal{L}, \Vdash)$ that properly extends $\mathbf{L}$ in Δ is not $\sim$-paraconsistent.

The following observation shows that maximal $\sim$-paraconsistency with respect to $Fm(\mathcal{L}^{\sim})$ is a rather liberal notion of maximal paraconsistency.

Proposition 1 *Let* $\mathbf{L} = (\mathcal{L}, \vdash)$ *be a* $\sim$*-paraconsistent logic that satisfies dni and dne, and let* $\rightarrow$ *be a proper implication for* $\mathbf{L}$. *Additionally we assume that* $\sim A, \sim B \nvdash A$ *for some formula* A *and a theorem* B *of* $\mathbf{L}$.[20] *Then* $\mathbf{L}$ *is maximally* $\sim$*-paraconsistent with respect to* $Fm(\mathcal{L}^{\sim})$.

Proof Suppose $(\mathcal{L}, \Vdash)$ properly extends $\mathbf{L}$ in $Fm(\mathcal{L}^{\sim})$. Let $(\{A_1, \ldots, A_n\}, A) \in \Vdash$ but $(\{A_1, \ldots, A_n\}, A) \notin \vdash$, with $A_1, \ldots, A_n, A \in Fm(\mathcal{L}^{\sim})$. Since $\Vdash$ satisfies Cut, dni and dne, we may assume without loss of generality that $A_1, \ldots, A_n, A$ are literals, i.e., sentence letters or negated sentence letters. We also may assume that $A_1, \ldots, A_n, A$ and the theorem B mentioned in the assumption have no common sentence letters. Since $\vdash$ satisfies Reflexivity and Monotonicity, A is different from each of the A_j $(1 \leq j \leq n)$. Suppose A is a sentence letter p. We show by reductio that $\sim p$ is not among $A_1, \ldots, A_n$. Suppose it is. Then $1 < n$ because $\sim$ is a proper negation. We substitute all the sentence letters different from p in $A_1, \ldots, A_n$ by the theorem B and apply Cut to $\Vdash B$ and $A_1, \ldots, \sim p, \ldots, A_n \Vdash p$, so as to remove B. Then we obtain $\sim p \Vdash p$ or $\sim p, \sim B \Vdash p$, quod non. Next, suppose A is a negated sentence letter $\sim p$ and p is among the $A_1, \ldots, A_n$. We substitute $\sim p$ for p and use dne and Cut to obtain $A_1, \ldots, \sim p, \ldots, A_n \Vdash p$, so that we can proceed as in the previous case. We thus have $A_1, \ldots, A_n \Vdash p$ or $A_1, \ldots, A_n \Vdash \sim p$, where neither p nor $\sim p$ is an element from $\{A_1, \ldots, A_n\}$. But then, given closure under uniform substitution, dni, dne, Cut, and Monotonicity, $\Vdash$ is not $\sim$-paraconsistent if there are literals $q, \sim q$ in $\{A_1, \ldots, A_n\}$, or $\Vdash$ is trivial and hence fails to be $\sim$-paraconsistent. □

[20] Clearly, $\sim$ is a proper negation for $\mathbf{L}$ under our assumptions.

Examples and counterexamples

- **N4** and **N4**$^{\perp}$ are maximally $\sim$-paraconsistent with respect to $Fm(\mathcal{L}^{\sim})$ because strong negation satisfies double negation introduction and elimination. Moreover, p does not follow from $\sim p$ and $\sim B$ in **N4** (**N4**$^{\perp}$) if p does not occur in B.
- For the same reason, every paraconsistent axiomatic extension of **N4**$^{\perp}$ and **N4** is maximally $\sim$-paraconsistent with respect to $Fm(\mathcal{L}^{\sim})$.
- Co-negation $\sim$ in dual-intuitionistic logic, bi-intuitionistic logic, and Priest-da Costa logic is paraconsistent but these systems are not maximally $\sim$-paraconsistent with respect to $Fm(\mathcal{L}^{\sim})$. The addition of weak *ex falso* $(\{\sim A, \sim\sim A\}, B)$ results in a logic intermediate between these logics and classical logic, see [26].

We now consider axiomatic extensions of propositional logics.

Definition 12 A two-place connective $\to$ from a propositional language $\mathcal{L}$ with a proper negation $\sim$ is a *connexive implication* for a propositional logic $(\mathcal{L}, \vdash)$ iff for every $A, B \in Fm(\mathcal{L})$ the following holds: $\sim(A \to B) \vdash (A \to \sim B)$ and $(A \to \sim B) \vdash \sim(A \to B)$.

Definition 13 Let $\mathbf{L} = (\mathcal{L}, \vdash)$ be a $\sim$-paraconsistent propositional logic and let $\Delta \subseteq Fm(\mathcal{L})$. A propositional logic $(\mathcal{L}, \Vdash)$ *properly a-extends* $\mathbf{L}$ *in the set* Δ iff $(\mathcal{L}, \Vdash)$ extends $\mathbf{L}$ in Δ (i.e., $(\vdash\restriction \Delta) \subseteq (\Vdash\restriction \Delta)$) and $(\{A \mid \varnothing \vdash A\} \restriction \Delta) \subset (\{A \mid \varnothing \Vdash A\} \restriction \Delta)$. The logic $\mathbf{L}$ is *a-maximally* $\sim$*-paraconsistent with respect to* Δ iff every propositional logic $(\mathcal{L}, \Vdash)$ that properly a-extends $\mathbf{L}$ in Δ is not $\sim$-paraconsistent.

Definition 14 An implication $(A \to B)$ is *left-literal* iff (i) A is a literal and (ii) every implication in B is left-literal. If $\mathcal{L}$ is a propositional language and $\Delta \subseteq Fm(\mathcal{L})$, then $Fm_l(\Delta) = \{A \in \Delta \mid A$ does not contain a subformula $(B \to C)$ that is not left-literal$\}$.

Definition 15 Let $\mathcal{L}$ be a propositional language. A formula $A \in Fm(\mathcal{L})$ is in *negation normal form (nnf)* iff it contains $\sim$ only in front of atomic subformulas.

Lemma 1 *Let* $\mathbf{L} = (\mathcal{L}, \vdash)$ *be a propositional logic that satisfies dni and dne, and let* $\to$ *be a connexive proper implication for* $\mathbf{L}$. *Then for every* $A \in Fm(\mathcal{L}^{\sim,\to})$ *there is a formula* $A' \in Fm(\mathcal{L}^{\sim,\to})$ *in nnf with* $A \vdash A'$ *and* $A' \vdash A$.

Proof By induction on the number n of connectives in $A \in Fm(\mathcal{L}^{\sim,\to})$. If $n = 0$ or $n = 1$, the claim is trivially true. Assume that $n + 1 \geq 2$ and that the claim holds for every $m \leq n$. If $A \equiv (B \to C)$, the claim follows by the induction hypothesis. If $A \equiv \sim\sim B$ the claim follows with dni and dne. Let $A \equiv \sim(B \to C)$ and assume that $B, \sim C$ are interderivable with formulas B' and C', respectively, in nnf. Then we have:

$$\dfrac{B' \vdash B \qquad \dfrac{\dfrac{\sim(B \to C) \vdash (B \to \sim C)}{\sim(B \to C), B \vdash \sim C} \qquad \sim C \vdash C'}{\sim(B \to C), B \vdash C'}}{\dfrac{\sim(B \to C), B' \vdash C'}{\sim(B \to C) \vdash (B' \to C')}}$$

$$\dfrac{B \vdash B' \qquad \dfrac{\dfrac{\dfrac{(B' \to C') \vdash (B' \to C')}{(B' \to C'), B' \vdash C'} \qquad C' \vdash {\sim}C}{(B' \to C'), B' \vdash {\sim}C}}{}}{\dfrac{\dfrac{(B' \to C'), B \vdash {\sim}C}{(B' \to C') \vdash (B \to {\sim}C)} \qquad (B \to {\sim}C) \vdash {\sim}(B \to C)}{(B' \to C') \vdash {\sim}(B \to C)}}$$

□

Proposition 2 *Let* $\mathbf{L} = (\mathcal{L}, \vdash)$ *be a* ${\sim}$*-paraconsistent logic that satisfies dni and dne, and let* $\to$ *be a connexive proper implication for* $\mathbf{L}$*. Additionally we assume that* ${\sim}A, {\sim}B \nvdash A$ *for some formula* A *and a theorem* B *of* $\mathbf{L}$*. Then* $\mathbf{L}$ *is a-maximally* ${\sim}$*-paraconsistent with respect to* $Fm_l(\mathcal{L}^{\sim,\to})$.

Proof Suppose that $(\mathcal{L}, \Vdash)$ is a proper a-extension of $\mathbf{L}$ in $Fm_l(\mathcal{L}^{\sim,\to})$. Let $\varnothing \Vdash A$ but $\varnothing \nvdash A$, with $A \in Fm_l(\mathcal{L}^{\sim,\to})$. If A is a literal, then $\Vdash$ is trivial and hence not paraconsistent. If A is not a literal, then by Lemma 1, A may be assumed to be an implication $(B_1 \to (B_2 \to \ldots (B_m \to B)\ldots))$ in nnf. Since $\to$ is a proper implication and $A \in Fm_l(\mathcal{L}^{\sim,\to})$, $B_1, \ldots, B_m \Vdash B$, where B and all the $B_1, \ldots, B_m$ are literals, whereas $B_1, \ldots, B_m \nvdash B$. We may now argue as in the proof of Proposition 1. □

Definition 16 Let $\mathbf{L} = (\mathcal{L}, \vdash)$ be a ${\sim}$-paraconsistent propositional logic and let $\Delta \subseteq Fm(\mathcal{L})$. A propositional logic $(\mathcal{L}, \Vdash)$ *properly l-extends* $\mathbf{L}$ *in* Δ iff $(\mathcal{L}, \Vdash)$ extends $\mathbf{L}$ in Δ (i.e., $(\vdash\restriction\Delta) \subseteq (\Vdash\restriction\Delta)$) and for some set of literals Θ, $(\{A \mid \Theta \vdash A\} \restriction \Delta) \subset (\{A \mid \Theta \Vdash A\} \restriction \Delta)$. The logic $\mathbf{L}$ is *l-maximally* ${\sim}$*-paraconsistent with respect to* Δ iff every propositional logic $(\mathcal{L}, \Vdash)$ that properly l-extends $\mathbf{L}$ in Δ is not ${\sim}$-paraconsistent.

The following observation immediately follows from Proposition 2.

Corollary 1 *Let* $\mathbf{L} = (\mathcal{L}, \vdash)$ *be a* ${\sim}$*-paraconsistent logic that satisfies dni and dne, and let* $\to$ *be a connexive proper implication for* $\mathbf{L}$*. Moreover, assume that* ${\sim}A, {\sim}B \nvdash A$ *for some formula* A *and a theorem* B *of* $\mathbf{L}$*. Then* $\mathbf{L}$ *is l-maximally* ${\sim}$*-paraconsistent with respect to* $Fm_l(\mathcal{L}^{\sim,\to})$.

Definition 17 Let $\mathbf{L} = (\mathcal{L}, \vdash)$ be a propositional logic. A binary connective $\wedge$ from $\mathcal{L}$ is a *proper conjunction* for $\mathbf{L}$ iff for every $A, B \in Fm(\mathcal{L})$ and every $\Delta \subseteq Fm(\mathcal{L})$ the following holds: ($\Delta \vdash A$ and $\Delta \vdash B$) iff $\Delta \vdash (A \wedge B)$. A binary connective $\vee$ from $\mathcal{L}$ is a *proper disjunction* for $\mathbf{L}$ iff for every $A, B \in Fm(\mathcal{L})$ and every $\Delta \subseteq Fm(\mathcal{L})$ the following holds: ($\Delta, A \vdash C$ and $\Delta, B \vdash C$) iff $\Delta, (A \vee B) \vdash C$.

The above equivalences are the corresponding "double-line rules" for conjunction and disjunction (restricted to single succedents) from [22]. Familiar left rules for conjunction and right rules for disjunction are then derivable:

$$\dfrac{\dfrac{(A \wedge B) \vdash (A \wedge B)}{(A \wedge B) \vdash A} \qquad \dfrac{\dfrac{(A \wedge B) \vdash (A \wedge B)}{(A \wedge B) \vdash B} \quad \dfrac{A \vdash C}{A, B \vdash C}}{A, (A \wedge B) \vdash C}}{\dfrac{(A \wedge B), (A \wedge B) \vdash C}{(A \wedge B) \vdash C}}$$

$$\dfrac{\dfrac{(A \wedge B) \vdash (A \wedge B)}{(A \wedge B) \vdash A} \qquad \dfrac{\dfrac{(A \wedge B) \vdash (A \wedge B)}{(A \wedge B) \vdash B} \quad \dfrac{B \vdash C}{A, B \vdash C}}{A, (A \wedge B) \vdash C}}{\dfrac{(A \wedge B), (A \wedge B) \vdash C}{(A \wedge B) \vdash C}}$$

$$\dfrac{\Delta \vdash A \quad \dfrac{(A \vee B) \vdash (A \vee B)}{A \vdash (A \vee B)}}{\Delta \vdash (A \vee B)} \qquad \dfrac{\Delta \vdash B \quad \dfrac{(A \vee B) \vdash (A \vee B)}{B \vdash (A \vee B)}}{\Delta \vdash (A \vee B)}$$

Definition 18 A proper disjunction $\vee$ for a propositional logic $\mathbf{L} = (\mathcal{L}, \vdash)$ is *Harropian* for $\mathbf{L}$ iff for every $A, B \in Fm(\mathcal{L})$ and every $\Delta \subseteq Fm(\mathcal{L})$ such that the formulas in Δ do not contain a disjunction as a strictly positive part[21] the following holds: $\Delta \vdash (A \vee B)$ iff ($\Delta \vdash A$ or $\Delta \vdash B$).

Definition 19 A proper conjunction $\wedge$ and a proper disjunction $\vee$ from a propositional language $\mathcal{L}$ with a proper negation $\sim$ are a *De Morgan conjunction*, respectively a *De Morgan disjunction* for a propositional logic $(\mathcal{L}, \vdash)$ iff for every $A, B \in Fm(\mathcal{L})$ the following holds: ${\sim}(A \wedge B) \vdash ({\sim}A \vee {\sim}B)$, $({\sim}A \vee {\sim}B) \vdash {\sim}(A \wedge B)$, ${\sim}(A \vee B) \vdash ({\sim}A \wedge {\sim}B)$, $({\sim}A \wedge {\sim}B) \vdash {\sim}(A \vee B)$.

Lemma 2 *Let* $\mathbf{L} = (\mathcal{L}, \vdash)$ *be a propositional logic that satisfies dni and dne, let* $\rightarrow$ *be a connexive proper implication for* **L**, *and let* $\wedge$ *and* $\vee$ *be a proper De Morgan conjunction, respectively a proper De Morgan disjunction for* **L**. *Then for every* $A \in Fm(\mathcal{L}^{\sim,\rightarrow,\wedge,\vee})$ *there is a formula* $A \in Fm(\mathcal{L}^{\sim,\rightarrow,\wedge,\vee})$ *in nnf with* $A \vdash A'$ *and* $A' \vdash A$.

Proof Analogous to the proof of Lemma 1. □

Proposition 3 *Let* $\mathbf{L} = (\mathcal{L}, \vdash)$ *be a* $\sim$*-paraconsistent logic that satisfies dni and dne, let* $\rightarrow$ *be a connexive proper implication for* **L**, *let* $\wedge$ *be a proper De Morgan conjunction for* **L** *and let* $\vee$ *be a proper Harropian disjunction for* **L**. *Additionally we assume that* ${\sim}A, {\sim}B \nvdash A$ *for some formula* A *and a theorem* B *of* **L**. *Then* **L** *is l-maximally* $\sim$*-paraconsistent with respect to* $Fm_l(\mathcal{L}^{\sim,\rightarrow,\wedge,\vee})$.

Proof Suppose that $(\mathcal{L}, \Vdash)$ is a proper l-extension of **L** in $Fm_l(\mathcal{L}^{\sim,\rightarrow,\wedge,\vee})$. Let Θ be a set of literals and let $\Theta \Vdash A$ but $\Theta \nvdash A$, with $A \in Fm_l(\mathcal{L}^{\sim,\rightarrow,\wedge,\vee})$. If A is a literal,

[21] That is, the disjunction is not contained in the antecedent of an implication.

then we may argue as in the proof of Proposition 1. If A is not a literal, then by Lemma 2, A may be assumed to be in nnf. We show by induction on the number n of binary connectives in A that if $\Theta \Vdash A$ and $\Theta \nVdash A$, then there exist literals $B_1, \ldots, B_n, B$ with $\Theta, B_1, \ldots, B_n \Vdash B$ and $\Theta, B_1, \ldots, B_n \nVdash B$ or with $\Theta \Vdash B$ and $\Theta \nVdash B$, so that we may argue as in the proof of Proposition 1. Assume $n = 1$. If A is an implication $(B_1 \to B_2)$, we have $\Theta, B_1 \Vdash B_2$; $\Theta, B_1 \nVdash B_2$. If A is a conjunction $(B_1 \wedge B_2)$, we have $\Theta \Vdash B_1$ and $\Theta \Vdash B_2$. Since, moreover, (i) $\Theta \nVdash B_1$ or (ii) $\Theta \nVdash B_2$, there is a case in which we may argue as in the proof of Proposition 1. If A is a disjunction $(B_1 \vee B_2)$, we have $\Theta \Vdash B_1$ or $\Theta \Vdash B_2$. Since, moreover, (i) $\Theta \nVdash B_1$ and (ii) $\Theta \nVdash B_2$ we again have a case in which we may argue as in the proof of Proposition 1. If $n = m + 1$, we may just use the induction hypothesis. If A is an implication, the induction hypothesis may be applied because of the restriction to left-literal implications. □

Examples

- All paraconsistent extensions of the connexive propositional logic **C** [35, 69, 70] are l-maximally $\sim$-paraconsistent with respect to $Fm_l(\mathcal{L}^{\sim,\to,\wedge,\vee})$ $(= Fm_l(\mathcal{L}))$.[22]

We argued that since paraconsistency is defined in terms of negation and derivability, it seems justified to require maximal paraconsistency with respect to $\mathcal{L}^{\sim}$. In view of this consideration and the above observation that, for example, **N4** and **N4**$^{\perp}$ are maximally $\sim$-paraconsistent with respect to $Fm(\mathcal{L}^{\sim})$, we propose the following definition as a still very tentative characterization of $\sim$-paraconsistent logics which is perhaps not "ideal" but "desirable".

Definition 20 A $\sim$-paraconsistent logic $\mathbf{L} = (\mathcal{L}, \vdash)$ is *desirable* iff it is maximally $\sim$-paraconsistent with respect to $Fm(\mathcal{L}^{\sim})$ and $\mathcal{L}$ contains a proper implication.

6 Another Conception of Maximality

Maximal paraconsistency is a kind of *minimal avoidance*; the idea is to have a logic that comprises as many inferences as possible but still admits non-trivial inconsistent theories. On this approach it is unproblematic to draw consequences from inconsistent premise sets as long as the logic is not "explosive". Another approach to paraconsistency aims at *maximal avoidance*: nothing at all follows from an inconsistent premise set, *ex contradictione nihil sequitur*, see [64]. If a logic in a language with negation $\sim$ has theorems, then satisfying *ex contradictione nihil sequitur* results in a system of non-monotonic logic because the simultaneous addition of premises A and $\sim A$ is precluded. In our view *maximal avoidance* is problematic. If there are premises, no matter whether in the case of finitary logics they are aggregated

[22] The system **C** is a non-trivial inconsistent system; $\sim(p \vee \sim p) \to (p \vee \sim p)$ and $\sim(\sim(p \vee \sim p) \to (p \vee \sim p))$, for example, are both provable. Therefore, one may wonder whether $\sim$ deserves to be viewed as a negation. Note that $\sim$ in **C** not only is a proper negation as defined above, but also has several negation-related properties; it satisfies the double negation and the De Morgan laws.

conjunctively by additive conjunction, $\wedge$, or by multiplicative conjunction, $\circ$, the premises *do* provide information. The premises A and $\sim A$ provide the information that A and the information that $\sim A$; the assumption $A \wedge B$ gives the information that A, that B, and that $A \wedge B$, and the assumption $A \circ B$ provides the information that $A \circ B$. From the perspective of information processing, multiple premises always do provide information.

Nevertheless, the idea that a contradiction provides no information has its attraction if one thinks of negation as a cancellation of propositional content, cf. [54]. But even if $\sim p$ cancels the content of p, an inconsistent theory $\{p, \sim p, q\}$ still provides the information that q. The motivation of paraconsistency by considerations of information processing does not support *ex contradictione nihil sequitur* for inconsistent theories. Note that q *does* stand in the strong Rescher–Manor consequence relation with (is a classical consequence of every maximally consistent subset of) the premise set $\{p, \sim p, q\}$ [59], but that relation is not reflexive.

The following variation of the concept of maximality is also possible. In the previous section we restricted the consequence relation to a set of formulas of a special form. Equally, we may restrict attention to non-trivial sets of premises. A set Δ of premises is non-trivial with respect to the consequence relation $\vdash$ if the set $\{A \mid \Delta \vdash A\}$ differs from the set of all formulas. Having in mind one or another reference logic with consequence relation $\vdash$, we can try to construct its paraconsistent variant whose consequence relation is as close as possible to $\vdash$ on non-trivial sets of premises. In fact, this approach is realized in the already mentioned adaptive logics, which are constructed so that the adaptive consequence relation coincides with the consequence relation of the upper limit logic on normal (equivalently, non-trivial) sets of premises. But this is done at the expense of non-monotonicity of adaptive consequence.

7 Another Methodology: Priest and Routley

In their presentation of systems of paraconsistent logic, Priest and Routley [57][23] apply a methodology that is considered here because it is quite different from the da Costa tradition of preserving as much as possible from classical logic, although they remark concerning their favoured approach to paraconsistency that "a pleasing feature of the semantics is that the set of zero degree logical truths is exactly the set of classical tautologies" (p. 169). This remark reflects Priest's [52, p. 235] methodological maxim, MM:

> Unless we have specific grounds for believing that paradoxical sentences are occurring in our argument, we can allow ourselves to use both valid and quasi-valid inferences

where an inference is understood to be quasi-valid if it is "valid provided all the truth values involved are classical (i.e., true only or false only)" [52, p. 231]. Clearly, there

[23] As Graham Priest told us, the paper was written already around 1979.

is a difference between accepting classical logic in specific contexts, say for languages without self-reference or for reasoning about decidable predicates, versus maximally approaching some reference logic, be it classical logic or some non-classical system. Graham Priest (personal communication) never advocated maximal paraconsistency.

Priest and Routley [57] identify three main approaches to paraconsistency and use certain desiderata to determine "the most satisfactory" approach (p. 180). In discussing the three approaches, they develop a number of "suitability requirements" (p. 175) any satisfactory logic and hence any satisfactory paraconsistent logic ought to satisfy. These are:

1. The availability of a "genuine" conjunction operation $\wedge$ that is adjunctive: $\{A, B\} \models A \wedge B$ (p. 159);
2. Recursive truth conditions (p. 159, p. 163);
3. The "normal relationships between conjunction, disjunction and negation," namely the De Morgan laws (p. 159);
4. The presence of valid multi–premise inferences (p. 161), which echoes in a sense Jaśkowski's requirement of being rich enough to enable practical inference;
5. The semantic evaluation clauses should be well-motivated (p. 163);
6. The presence of a negation that is "really negation" (p. 164 f.), which according to Priest and Routley means (i) the presence of a contradiction forming functor, so that they are inclined to require the law of non-contradiction to be valid, although they "do not wish to be too dogmatic about this" (p. 164), and (ii) that at least some of the "inferential properties traditionally associated with negation" hold;
7. The presence of an implication that satisfies *modus ponens* because "[n]o operator which fails to satisfy this can be implication" (p. 171);
8. Relevance: "an implication should hold between A and B only in virtue of some common content between A and B," where Anderson and Belnap's variable-sharing property provides a test for irrelevance (p. 171)[24];
9. The invalidity of Absorption, $(A \rightarrow (A \rightarrow B)) \rightarrow (A \rightarrow B)$, since any logic that comprises both *modus ponens* and Absorption gives rise to the Curry paradox (p. 172);
10. [not only $\{A, \sim A\} \models B$ and $A \wedge \sim A \models B$ should fail, but also] $A \equiv \sim A \models B$ should fail for an equivalence connective $\equiv$ (p. 173);
11. The validity of certain "natural implication rules", including $\{A \rightarrow B, A \rightarrow \sim B\} \models \sim A$ and the following form of contraposition, $A \rightarrow B \models \sim B \rightarrow \sim A$ (p. 174);
12. Transitivity of implication because "it seems to be such a fundamental principle of implication, almost as fundamental as *modus ponens*, that it should be given

[24]Priest and Routley explain that there are paraconsistent logics that are not relevance logics and that there exist relevance logics that are not paraconsistent. For the latter they refer to Ackermann's system of strong implication, which uses Disjunctive Syllogism in rule form (p. 182). However, they also emphasize that "any relevant logic will avoid the paraconsistently execrable *ex falso quodlibet* and therefore will be a *prima facie* candidate for a paraconsistent logic" (p. 177).

up only under the most extreme circumstances. Since there are other approaches which validate transitivity, these circumstances do not obtain" (p. 178).

This methodology is much more informal than the one developed by Arieli, Avron, and Zamansky, and hence it is more difficult to apply it. Several of the above conditions are completely unproblematic. It is, e.g., obviously uncontroversial to require that the semantic evaluation clauses ought to be well-motivated (but it would be interesting to further specify this extremely vague requirement). Some items from the above list of conditions, however, invite comments or objections.

As to recursive truth conditions, it is normally taken for granted that the set of atomic formulas forms the inductive basis. If positive and negative information are treated on a par, and if support of truth and support of falsity of atomic propositions are treated as independent notions in their own right so that in a four-valued setting both gaps and gluts of ordinary truth values are admitted, then it is quite natural to use the set of literals as the inductive basis for a recursive definition of the truth and falsity conditions of compound formulas. Under this conception, the separate truth and falsity conditions of complex formulas in the paraconsistent logics **N4**, $\mathbf{N4}^{\perp}$, and **C** are recursive.

As to negation, what negation really is, and whether there exists *the unique* and correct notion of negation is contentious, cf. [29, 31, 66]. To require that at least some of the inferential properties traditionally associated with negation hold seems, however, quite agreeable, in particular the De Morgan laws form a core set of negation principles if negation is to be understood as expressing falsity. In their concluding paragraph, Priest and Routley [57, p. 180] list contraposition as one of the right properties of negation. Requiring contraposition as a rule or as a valid schema is objectionable from the point of view of information processing with support of truth and support of falsity as independent notions. Suppose the implication connective satisfies the deduction theorem, as required above, so that $A \rightarrow B$ is provable iff B is derivable from A. If the information that A gives the information that B, it is not at all clear or obvious that the information that $\sim B$ provides the information that $\sim A$.[25] If negation expresses falsity, in a four valued setting it may happen that an information state supports the truth of A and B, supports the falsity of B, but fails to support the falsity of A.

Relevance and the invalidity of Absorption are general suitability requirements and are imposed with no particular view on paraconsistency; relevance is imposed to "get implication right"(p. 172). Priest and Routley present what they take to be the most satisfactory approach to paraconsistency in the context of a classification of relevance logics ("relevant logics," as they call them) and advocate paraconsistent depth relevance logics[26]:

[25] An anonymous referee remarked that using "giving information" to interpret the consequence relation is at odds with using it to interpret the implication connective in a discussion of the contraposition schema, but contraposition as a schema and *modus ponens* give one the problematic contraposition rule.

[26] Note that in the statement of Disjunctive Syllogism, $(A \wedge (\sim A \vee B)) \rightarrow B$, in [57, p. 156] there is a disturbing typographical mistake: $(A \wedge (\sim A \wedge B)) \rightarrow B$.

Accept Disjunctive Syllogism ↗ Retain Transitivity ↗ Connexive positions
Accept Disjunctive Syllogism ↗ Retain Transitivity ↘ Conceptivist (Parry) systems
Accept Disjunctive Syllogism ↘ Reject Transitivity ⟶ Sieve systems
Reject Disjunctive Syllogism ↗ Retain Absorption ⟶ Anderson-Belnap systems
Reject Disjunctive Syllogism ↘ Reject Absorption ⟶ Depth relevant logics

The notion of a depth relevance logic is due to Ross Brady [16], who defines depth relevance as a strengthened relevance condition. Intuitively, the depth of an occurrence of a subformula in a given formula is the number of implications under which the occurrence of the subformula is nested. The depth of the occurrence of p in $r \to ((p \to q)) \to q)$, for instance, is 3. The depth relevance condition holds for a logic **L** iff for all formulas A and B, the provability of $A \to B$ in **L** implies that for some sentence letter p and natural number d, there exists an occurrence of p in A at depth d and also an occurrence of p in B at depth d. An example of a formula the provability of which violates depth relevance is $(p \wedge (p \to q)) \to q$, so that it is clear that the well-known relevance logics **E**, **T**, and **R** do not enjoy depth relevance, and Brady [16] remarks that the addition of Absorption to the weak, depth relevance logic **B** results in a loss of depth relevance.

The above schema needs an amendment. The placement of connexive positions in the schema is based on the connexive logics that had been dealt with in literature when [57] was written. Priest and Routley [57, p. 181] argue that

> [s]ince in connexive logics A and $\sim A$ cancel one another, A and $\sim A$ are never designated together, and $A \wedge \sim A$ is not designated. Thus both $\{A, \sim A\} \models B$ and $A \wedge \sim A \models B$ hold (on designation-preserving accounts), and connexive logics are not paraconsistent.[27]

In the meantime, however, paraconsistent connexive logics have been presented. As a result, the use of a connexive implication is an option also in relevance (relevant) logics that retain or reject Absorption. Therefore, the lower part of the above figure can be expanded to the following diagram:

Reject DS ↗ Retain Absorption ⟶ A-B systems ↗ Connexive implication
Reject DS ↗ Retain Absorption ⟶ A-B systems ↘ Conjunctive negated implication
Reject DS ↘ Reject Absorption ⟶ Depth rel. logics ↗ Connexive implication
Reject DS ↘ Reject Absorption ⟶ Depth rel. logics ↘ Conjunctive negated implication

---

[27] As to Parry systems, also called "containment logics" or "logics of analytic implication", they explain that "on the so far received semantics for these systems, A and $\sim A$ are never designated together, and $A \wedge \sim A$ is not designated" (p. 181). A recent reference to Parry systems is [27].

where "conjunctive negated implication" refers to interpreting $\sim(A \to B)$ as $A \wedge \sim B$.

A linear, contraction-free relevance logic with an orthodox understanding of strongly negated implications has been presented in [65]. A linear relevance logic with a connexive implication can be obtained from the cut-free sequent calculus for the connexive logic **C** in [35] by deleting the structural rules of weakening and contraction. Priest and Routley's methodology for developing systems of paraconsistent logic thus seems to lead to linear logics that contain either an implication that is in accordance with the conjunctive understanding of negated implications or a connexive implication (if one assumes that no other readings of negated implications are convincing and, moreover, only systems with one implication and one negation are countenanced).

8 Minimality Instead of Maximality

In this section we sketch an approach to constructing paraconsistent logics that is based on principles essentially different from the maximality approach. It assumes a variety of different reference logics, agrees well with an informational treatment of paraconsistency, and is motivated by David Nelson's work [45] devoted to the separation of concepts. Earlier Nelson had suggested a constructive logic with strong negation [44] as a new version of intuitionistic logic. According to [45], the most important motivation for constructing this logic was the trivialization of the constructive meaning of negated formulas in intuitionistic arithmetic: "Under the recursive interpretation of a formal system for intuitionistic arithmetic, the provable implications of the form $A \supset 1 = 0$ receive a trivial interpretation". Further justification for a logic with strong negation can be obtained in terms of concept separation [45, p. 215]:

> As we have suggested earlier, an argument favoring intuitionistic logic over the classical is the fact that the intuitionistic logic allows the classical distinctions in meaning and further ones besides. Classical logic is open to possible objection in that it identifies certain constructively distinct entities. Since we are speaking here of formal systems, we are interested in the general question of finding when one formal system allows distinctions among concepts which are not possible in the other. This involves the general question of methods of representing one system in another.

The answer to the mentioned question is the following.

Definition 21 ([45]) Let $\mathbf{L}_2$ be a subsystem of $\mathbf{L}_1$, let $\equiv_i$ be a specified equivalence symbol of $\mathbf{L}_i$. A transformation * of formulas of $\mathbf{L}_1$ to formulas of $\mathbf{L}_2$ is said to be regular just in case:

1. For every formula A of $\mathbf{L}_1$, $A \equiv_1 A^*$ is provable in $\mathbf{L}_1$.
2. For every pair of formulas A and B of $\mathbf{L}_1$, if $A^* \equiv_2 B^*$ is provable in $\mathbf{L}_2$, then $A \equiv_1 B$ is provable in $\mathbf{L}_1$.
3. If E is an atomic formula, then E^* is E.

According to [45], if there is a regular transformation from $\mathbf{L}_1$ to $\mathbf{L}_2$, then $\equiv_2$ in $\mathbf{L}_2$ allows all the distinctions among concepts which are allowed by $\equiv_1$ in $\mathbf{L}_1$. If there is no regular transformation from $\mathbf{L}_1$ to $\mathbf{L}_2$, then $\equiv_1$ allows distinctions which are not regularly presented by $\equiv_2$ in $\mathbf{L}_2$.

In [44, § 5], it was proved that in **QN3**, the first-order version of the three-valued (explosive) Nelson logic, the strong negation is independent from the other connectives. From this result one can infer that there is no regular transformation from **QN3** to its intuitionistic subsystem. This means that first-order explosive Nelson logic distinguishes concepts better than first-order intuitionistic logic.

If we have any explosive logic $\mathbf{L}_1$ and an arbitrary paraconsistent sublogic $\mathbf{L}_2$ in the same language, then it is clear that the identical transformation is a regular transformation from $\mathbf{L}_1$ to $\mathbf{L}_2$. Thus, every paraconsistent subsystem allows all the distinctions among concepts which are allowed in its explosive extension. Moreover, the equivalence connective of a paraconsistent logic allows to distinguish contradictions, which are equivalent in explosive logics. It makes sense, however, to strengthen the definition by Nelson, to show that in passing from an explosive logic to a paraconsistent subsystem we do not lose the expressive power of the reference logic. We also have to reject the condition 3. For Nelson this condition means that "...both systems are concerned with the same subject matter and start with the same basic concepts." However, if we pass to a paraconsistent logic, we have to take into consideration new concepts corresponding to non-equivalent contradictions as well as concepts which are logically incomparable with contradictory concepts. So, a transformation of atomic formulas must be non-trivial and must distinguish the concepts treated by the explosive reference logic among an extended "universe" of concepts taken into consideration by a paraconsistent version of this logic.

Further, a paraconsistent subsystem must have essentially the same non-negative connectives. Therefore, it is natural to assume that the desired transformation commutes with all these connectives. Moreover, it is natural to require that a paraconsistent logic and its explosive extension have the same negation-free fragment.

Definition 22 Let $\mathbf{L}_2$ be a paraconsistent subsystem of an explosive logic $\mathbf{L}_1$. A transformation * from the language of $\mathbf{L}_1$ into itself is *essential* iff it commutes with every connective except from negation.

The logic $\mathbf{L}_2$ is called an *expressive paraconsistent subsystem* of $\mathbf{L}_1$ iff there is an essential transformation that faithfully embeds $\mathbf{L}_1$ into $\mathbf{L}_2$ and both logics $\mathbf{L}_1$ and $\mathbf{L}_2$ have the same negation-free fragment.

It is known (see [47]) that the minimal logic of Johansson is an expressive paraconsistent subsystem of intuitionistic logic, whereas the version $\mathbf{N4}^{\perp}$ with intuitionistic negation of Nelson paraconsistent logic is an expressive paraconsistent subsystem of the explosive Nelson logic **N3**. For example, in the case of minimal logic the required transformation * can be defined as follows: A^* is obtained from A by replacing every occurrence of a sentence letter p by $p \vee \bot$. The atomic formulas p and $\bot$ are incomparable wrt the consequence relation of minimal logic, whereas $p \vee \bot$ is a consequence of $\bot$.

The above definition imposes some minimality condition on the paraconsistent subsystem: we should not lose the expressive power of the explosive reference logic. To which extent the expressive power must increase is an open question.[28] It depends on the intended applications of the defined paraconsistent sublogic and the superstructures over it we want to consider. A more detailed answer to this question is the subject of further investigations. Here we give only one example: how to explicate the consequence relation between contradictory concepts in (axiomatic) extensions of minimal logic. With an arbitrary extension **L** of minimal logic we associate its negative counterpart $\mathbf{L}_{neg} = \mathbf{L} + \{\bot\}$ (see [47]). This logic is contradictory, but non-trivial if **L** is not explosive. Moreover, it turns out that the consequence relation between formulas in $\mathbf{L}_{neg}$ simulates the consequence relation in **L** between contradictions constructed from these formulas. More exactly, the transformation $C(A) = A \wedge \neg A$ defines a strong embedding of $\mathbf{L}_{neg}$ into **L**, it preserves not only the set of theorems, but the consequence relation too. Of course, we may need more sophisticated superstructures, e.g., belief revision systems transforming contradictory theories over a paraconsistent logic into consistent theories over an explosive reference logic, and so on.

9 Summary

We argued that classical logic should not be used as a reference logic for developing systems of paraconsistent logic. Moreover, we suggested to relativize a certain notion of maximal paraconsistency to a given subset of the set of all formulas, in particular to the set $Fm(\mathcal{L}^{\sim})$ of all formulas containing at most negation as a logical connective. We also argued that the guiding motivation for the development of systems of paraconsistent logic should be neither epistemological nor ontological, but informational and pointed to problematic aspects of *ex contradictione nihil sequitur* as another conception of maximality. Finally, we commented on Priest and Routley's methodology that does not make use of a notion of maximal paraconsistency. The combination of the latter methodology and the requirement of maximal $\sim$-paraconsistency with respect to $Fm(\mathcal{L}^{\sim})$ seems to lead to a class of "ideal" paraconsistent logics that comprises the contraction-free fragments of the logics **N4**, $\mathbf{N4}^{\bot}$, and **C**.

There are many directions for future research. These include the presentation of other examples of logics that are maximally paraconsistent with respect to certain fragments of their language, a rigorous formal development of (part of) Sylvan and Priest's suitability requirements, a further development of the minimality approach

[28] The fact that requiring an increase of expressive power is a natural demand justifies regarding the above definition as imposing a minimality condition.

of the previous section, and a discussion of whether one should at all use a reference logic for the development of paraconsistent logics.

Acknowledgments We would like to thank the audiences at *Paraconsistent Reasoning in Science and Mathematics*, Munich, June 2014 and at the New York Philosophical Logic Group meeting on September 8, 2014 for various comments. In particular we are grateful to Graham Priest, João Marcos, and Itala D'Ottaviano. Also, we would like to thank Ofer Arieli, Arnon Avron, and Anna Zamansky for their detailed comments and for making available to us the chapter on negation and paraconsistency of a forthcoming book. We are also grateful to two anonymous referees and to Yaroslav Shramko and Jc Beall for their comments on an earlier draft of the present paper. We take sole responsibility for all remaining shortcomings.

References

1. Aberdein, A., & Read, S. (2009). "The philosophy of alternative logics". In L. Haaparanta (Ed.), *The Development of Modern Logic* (pp. 613–723). Oxford: Oxford UP.
2. Adriaans, P. "Information". In E.N. Zalta (Ed.), *The Stanford Encyclopedia of Philosophy* (Fall 2013 Edition). http://plato.stanford.edu/archives/fall2013/entries/information/.
3. Almukdad, A., & Nelson, D. (1984). Constructible falsity and inexact predicates. *Journal of Symbolic Logic*, *49*, 231–233.
4. Anderson, A.R., Belnap, N.D. (1975) *Entailment: The logic of relevance and necessity (Vol. I)*, Princeton: Princeton University Press.
5. Arieli, O., Avron, A., & Zamansky, A. (2011). What is an ideal logic for reasoning with inconsistency? *Proceedings IJCAI*, *2011*, 706–711.
6. Arieli, O., Avron, A., & Zamansky, A. (2011b). Maximal and premaximal paraconsistency in the framework of three-valued semantics. *Studia Logica*, *97*, 31–60.
7. Arieli, Ofer, Avron, Arnon, & Zamansky, Anna. (2011c). Ideal paraconsistent logics. *Studia Logica*, *99*, 31–60.
8. Avron, A. (1991). Natural 3-valued logics: Characterization and proof theory. *Journal of Symbolic Logic*, *56*, 276–294.
9. Avron, A. (1999). "Negation: Two points of view". In D. Gabbay & H. Wansing (Eds.), *What is Negation?* (pp. 3–22). Dordrecht: Kluwer Academic Publishers.
10. Avron, A. (2002). "On negation, completeness and consistency". In D. Gabbay & F. Guenthner (Eds.), *Handbook of Philosophical Logic*, 2nd edn, vol. 9, pp. 287–319. Dordrecht: Kluwer Academic Publishers.
11. Avron, A. (2014). The Classical Constraint on Relevance. *Log. Univers.*, *8*, 1–15.
12. Barwise, J. (1988). Three views of common knowledge. In Moshe Y. Vardi (Ed.), *Proceedings of TARK 2* (pp. 365–397). San Francisco: Morgan Kaufmann Publishers.
13. Batens, D. (1980). Paraconsistent extensional propositional logics. *Logique et Analyse*, *90*(91), 195–234.
14. Batens, D. (2013). "New Arguments for Adaptive Logics as Unifying Frame for the Defeasible Handling of Inconsistency". In K. Tanaka, F. Berto, E. Mares, & F. Paoli (Eds.), *Paraconsistency: Logic and Applications* (pp. 101–121). Dordrecht: Springer.
15. Beall, Jc. (2013). Free of Detachment: Logic, Rationality, and Gluts, *Noûs*, first published online. doi:10.1111/nous.12029.
16. Brady, R. T. (1984). Depth Relevance of Some Paraconsistent Logics. *Studia Logica*, *43*, 63–73.
17. Carnielli, W., Coniglio, M., & Marcos, J. (2007). "Logics of formal inconsistency". In D. Gabbay, F. Guenthner (Eds.), *Handbook of Philosophical Logic*, 2nd edn, vol. 14, 1–93. Dordrecht: Springer.

18. Carnielli, W., & Marcos, J. (2002). "A Taxonomy of C-systems". In W. Carnielli, M. Coniglio, & I. D'Ottaviano (Eds.), *Paraconsistency: The Logical Way to the Inconsistent* (pp. 1–94). New York: Marcel Dekker.
19. da Costa, N. (1974). On the theory of inconsistent formal systems. *Notre Dame Journal of Formal Logic, 15*, 497–510.
20. da Costa, N., & Alves, E. (1981). Relations between paraconsistent logic and many-valued logic. *Bulletin of the Section of Logic, 10*, 185–191.
21. Crabbé, M. (2011). Reassurance for the logic of paradox. *Review of Symbolic Logic, 4*, 479–485.
22. Došen, K. (1994). "Logical constants as punctuation marks", Notre Dame Journal of Formal Logic 30 (1989), 362–381; reprinted in a slightly amended version. In D. M. Gabbay (Ed.), *What is a Logical System?* (pp. 273–296). Oxford: Oxford University Press.
23. Dummett, M. (1978). *Truth and Other Enigmas*. London: Duckworth.
24. Dunn, J.M. (2001)."The Concept of Information and the Development of Modern Logic". In W. Stelzner, M. Stöckler (Eds.), *Zwischen traditioneller und moderner Logik. Nichtklassische Ansätze*, 423–447. Mentis: Paderborn.
25. Dunn, J.M. (2008). "Information in Computer Science". In P. Adriaans, J. van Benthem (Eds.), *Philosophy of Information*, Vol. 8 of the D. Gabbay, P. Thagard, J. Woods, (Eds.), *Handbook of The Philosophy of Science* 581–608. Amsterdam: North-Holland.
26. Ferguson, T. M. (2014). Extensions of Priest-da Costa logic. *Studia Logica, 102*, 145–174.
27. Ferguson, T.M. (2014). "Logics of Nonsense and Parry Systems", *Journal of Philosophical Logic*, published online. doi:10.1007/s10992-014-9321-y.
28. Floridi, L. (2015). "Semantic Conceptions of Information". In E.N. Zalta (Ed.), *The Stanford Encyclopedia of Philosophy* (Spring 2015 Edition). http://plato.stanford.edu/archives/spr2015/entries/information-semantic/.
29. Gabbay, D. M., & Wansing, H. (Eds.). (1999). *What is Negation?*. Dordrecht: Kluwer Academic Publishers.
30. Goré, R. (2000). Dual intuitionistic logic revisited. In R. Dyckhoff (Ed.), *Automated Reasoning with AnalyticTableaux and Related Methods, Springer Lecture Notes in AI 1847* (pp. 252–267). Berlin: Springer.
31. Horn, L.R. Wansing, H. (2015). "Negation". In E.Z. (Ed.), *The Stanford Encyclopedia of Philosophy* (Spring 2015 Edition). http://plato.stanford.edu/archives/spr2015/entries/negation/.
32. Humberstone, L. (2000). Contra-Classical Logic. *Australasian Journal of Philosophy, 78*, 438–474.
33. Jaśkowski, S. (1999). 'Rachunek zdań dla systemów dedukcyjnych sprzecznych", *Studia Societatis Scientiarum Torunensis, Sectio A, Vol. I*, No. 5, Toruń 1948, 57–77. English translation as "A propositional calculus for inconsistent deductive systems" *Logic and Logical Philosophy* 7, 35–56, cited after the English translation.
34. Johansson, I. (1936). Der Minimalkalkül, ein reduzierter intuitionistischer Formalismus. *Compositio Mathematica, 4*, 119–136.
35. Kamide, N. Wansing, H. (2011). "Connexive modal logic based on positive *S*4", In J.-Y. Bèziau, M. Coniglio (Eds.), *Logic without Frontiers. Festschrift for Walter Alexandre Carnielli on the Occasion of His 60th Birthday*, 389–409. London: College Publications.
36. Karpenko, A. S. (1999). Jaśkowski's criterion and three-valued paraconsistent logics. *Logic and Logical Philosophy, 7*, 81–86.
37. Ketonen, J., & Weyhrauch, R. W. (1984). A decidable fragment of predicate calculus. *Theoretical Computer Science, 32*, 297–307.
38. Komori, Y. (1986). Predicate logics without the structure rules. *Studia Logica, 45*, 393–404.
39. Lenzen, W. (1996). Necessary conditions for negation operators. In H. Wansing (Ed.), *Negation* (pp. 37–58). A Notion in Focus, Berlin: de Gruyter.
40. Lin, Z. (1995). "Paraconsistent Circumscription: First-Order Case", *Proceedings of the 25th International Symposium on Multiple-Valued Logic, ISMVL*, 1995, 112–116.
41. Marcos, J. (2005). On negation: Pure local rules. *Journal of Applied Logic, 3*, 185–219.
42. Mares, E. (2013). Information, Negation, and Paraconsistency. In K. Tanaka, F. Berto, E. Mares, & F. Paoli (Eds.), *Paraconsistency: Logic and Applications* (pp. 43–55). Dordrecht: Springer.

43. McCall, S. (2012). A History of Connexivity. In D. M. Gabbay, F. J. Pelletier, & J. Woods (Eds.), *Logic: A History of its Central Concepts* (Vol. 11, pp. 415–449)., Handbook of the History of Logic Amsterdam: Elsevier.
44. Nelson, D. (1949). Constructible Falsity. *Journal of Symbolic Logic*, *14*, 16–26.
45. Nelson, D. (1959). Negation and separation of concepts in constructive systems. In A. Heyting (Ed.), *Constructivity in Mathematics* (pp. 208–225). Amsterdam: North-Holland.
46. Odintsov, Sergei P. (2005). The class of extensions of Nelson paraconsistent logic. *Studia Logica*, *80*, 291–320.
47. Odintsov, S.P. (2008). *Constructive Negations and Paraconsistency*. Dordrecht: Springer.
48. Odintsov, S.P. & Wansing, H. (2003). "Inconsistency-tolerant Description Logic. Motivation and Basic Systems". In V. Hendricks, J. Malinowski (Eds.), *Trends in Logic. 50 Years of Studia Logica*, 301–335. Dordrecht: Kluwer Academic Publishers
49. Odintsov, S. P., & Wansing, H. (2008). Inconsistency-tolerant Description Logic. Part II: Tableau Algorithms. *Journal of Applied Logic*, *6*, 343–360.
50. D'Ottaviano, I. (1985). The completeness and compactness of a three-valued first-order logic. *Revista Colombiana de Matematicas*, *19*, 31–42.
51. Prawitz, D. (2015). "Explaining deductive inference", to appear. In H. Wansing (Ed.), *Dag Prawitz on Proofs and Meaning*. Dordrecht: Springer.
52. Priest, G. (1979). The logic of paradox. *Journal of Philosophical Logic*, *8*, 219–241.
53. Priest, G. (2006). "Minimally Inconsistent LP", *Studia Logica* 50 (1991), 321–331, reprinted with correction as chapter 16 of G. Priest, *In Contradiction*, 2nd edn., Oxford: Clarendon Press.
54. Priest, G. (1999). Negation as Cancellation, and Connexive Logic. *Topoi*, *18*, 141–148.
55. Priest, G. (2009). Dualising intuitionistic negation. *Principia*, *13*, 165–184.
56. Priest, G. (2011). First-order da Costa logic. *Studia Logica*, *97*, 183–198.
57. Priest, G., & Routley, R. (1989). Systems of paraconsistent logic. In G. Priest, R. Routley, & J. Norman (Eds.), *Paraconsistent Logic: Essays on the Inconsistent* (pp. 142–155). Munich: Philosophia Verlag.
58. Rauszer, C. (1980). *"An algebraic and Kripke-style approach to a certain extension of intuitionistic logic", Dissertationes Mathematicae 167*. Polish Academy of Sciences: Institute of Mathematics.
59. Rescher, N., & Manor, R. (1970). "On inference from inconsistent premises", *Theory and Decision* 1 (1970-71), 179–217.
60. Ripley, D. (2011). Negation, denial, and rejection. *Philosophy Compass*, *6*, 622–629.
61. Sette, A. M. (1973). On the propositional Calculus P^1. *Mathematica Japonicae*, *16*, 173–180.
62. Shramko, Y., & Wansing, H. (2011). *Truth and Falsehood. An Inquiry into Generalized Logical Values*, Trends in Logic, vol. 36, Dordrecht: Springer.
63. Straßer, C. (2014). *Adaptive Logics for Defeasible Reasoning, Trends in Logic* (Vol. 38). Dordrecht: Springer.
64. Wagner, G. (1991). "Ex contradictione nihil sequitur", In *Proceedings of IJCAI-91, 12th International Joint Conference on Artificial Intelligence*, Sydney, Australia, Morgan Kaufmann, San Francisco, California, 538–543.
65. Wansing, H. (1993). *The Logic of Information Structures* (Vol. 681)., Lecture Notes in AI Berlin: Springer.
66. Wansing, H. (Ed.). (1996). *Negation*. de Gruyter: A Notion in Focus, Berlin.
67. Wansing, H. (1999). "Contradiction and contrariety. Priest on negation". In J. Malinowski, A. Pietruszczak (Eds.), *Essays in Logic and Ontology*, 81–93. Amsterdam: Rodopi.
68. Wansing, H. (2001). "Negation". In L. Goble (Ed.), *The Blackwell Guide to Philosophical Logic* (pp. 415–436). Cambridge/MA: Basil Blackwell Publishers.
69. Wansing, H. (2005). "Connexive Modal Logic", In R. Schmidt et al. (Eds.), *Advances in Modal Logic. Volume 5*, King's College Publications, London, 367–383. http://www.aiml.net/volumes/volume5/.
70. Wansing, H. (2014). "Connexive Logic", In: E.N. Zalta (Ed.), *The Stanford Encyclopedia of Philosophy* (Fall 2014 Edition), first published 2006. http://plato.stanford.edu/archives/fall2014/entries/logic-connexive/.

71. Wansing, H. (2008). Constructive negation, implication, and co-implication. *Journal of Applied Non-Classical Logics*, *18*, 341–364.
72. Wansing, H. (2013). Falsification, natural deduction, and bi-intuitionistic logic. *Journal of Logic and Computation*, published online July 2013. doi:10.1093/logcom/ext035.
73. Weber, Z. (2010). "Paraconsistent Logic", In J. Fieser, B. Dowden (Eds.), *Internet Encyclopedia of Philosophy*.

Paraconsistent Computation and Dialetheic Machines

Zach Weber

Abstract Are there are any properly paraconsistent computations—effective procedures that are recognizable as such, but which are not recognized by non-paraconsistent logic? First we motivate a positive answer, from arguments by Sylvan and Copeland, Routley, and Priest. Then we look at some simple formulations of dialetheic machines and their basic properties, and discuss these in relation to the halting problem.

Keywords Dialetheism · Inconsistent computation

1 Introduction

1.1 More Things in Heaven and Earth ...

In inconsistent mathematics, some objects exist that cannot according to any other practice. For example, in a dialetheic naive set theory, e.g. [10, 22, 24] (in [4]), the collection of all non-self-membered sets is itself a set, and so is inconsistent, both self-membered and not. Famous diagonal arguments that would otherwise conclude in paradox are proofs that end in theorems. Closer to the ground, in what might be thought of as 'naive computability theory', it seems plausible that analogous diagonal arguments—like those around the famous halting problem [23, p. 24]—would lead to paradoxical theorems in the same way. If so, it makes sense to look for novel mathematical computational objects, analogous to inconsistent sets. Are there any *properly* paraconsistent effective procedures? Are there computations that are recognizable as such, but which are not recognized by non-paraconsistent logic? In the first half of this chapter, we motivate a positive answer to this question, in the spirit of [9, 29]; in the second half, we sketch some of the technical details of dialetheic machines.

Z. Weber (✉)
Department of Philosophy, University of Otago, Dunedin, New Zealand
e-mail: zach.weber@otago.ac.nz

H. Andreas and P. Verdée (eds.), *Logical Studies of Paraconsistent Reasoning in Science and Mathematics*, Trends in Logic 45,
DOI 10.1007/978-3-319-40220-8_13

1.2 ... and How to Talk About Them

Any discussion of computation will involve terms like 'algorithm', 'function', 'recursive', and 'machine'. And 'not', for that matter. Given that the paragraph above promises to countenance some highly non-classical notions, in highly unfamiliar settings, a word about the vocabulary and perspective of this chapter is needed before we get going.

To rethink the conceptual, logical, and mathematical foundations of computation, in light of (radical) alternative possibilities emerging from paraconsistent logic, requires us *not* to take terms to be pegged to classical logic/mathematics. For example, if 'consistency of a theory' just means 'not every sentence is a theorem' then the suggestion of an inconsistent theory becomes absurd. A fortiori the notion of an 'inconsistent function'. We have to stand at some remove from the orthodox understanding of terms, however innocuous they may seem. While the terms and the (meta)logic of the chapter are not definitively situated in one particular paraconsistent system (for the sake of generality), they are definitively *not* situated in a classical system.

The informal idea of a step-by-step program that can be executed with diligence but no creativity was codified in the 20th century as an algorithm. The broad idea, only begun here, is to return to Turing's original analysis, but to imagine, counterfactually, that Turing's philosophical reasoning was grounded in paraconsistency. Now, to conduct this thought experiment, we must also imagine a fair amount of arithmetic in the background; but the nature of arithmetic must itself be one of the moving parts in the research, not a fixed formal object. In the pages below, we will entertain alternative models of arithmetic, some of which are *finite*, some of which are *inconsistent* but in which still $0 = 1$ is not a theorem, and some of which have very different models of the successor 'function'. Defaulting to classical Peano Arithmetic would preempt some of the very decisions we want to reconsider, and would render the discussion incoherent.[1]

The discussion is oriented in the direction of a strong paraconsistency: dialetheism. There are many non-dialetheic approaches to paraconsistency, and so paraconsistent computation, some discussed below (e.g. Sect. 3.1). Our research question makes dialetheism—strong paraconsistency—central to the findings. A mathematics embedded in weaker-than-classical logic calls out for stronger-than-classical subject matter.

---

[1]With so much up in the air, then, it is not even possible to ask yet whether e.g. there are purely paraconsistent computations that make use of primitive recursive functions. The question isn't well-defined. Similarly, literature on the different formal definitions of algorithms [8] is too far along in its development to apply here. Perhaps for these reasons, the pioneering Copeland and Sylvan [29] talk about *algorithmic derivations*, not recursive functions. In that work, they also talk about dialetheic machines, which I have embraced for the title, with the understanding that machines are not attached to a particular formal meaning, and are mathematical objects, not empirical ones.

2 Motivation: Two Paradoxes

2.1 *The Paradox of All Algorithmic Functions*

An abstract background motivation for looking to paraconsistency is a putative paradox in the foundations of recursion theory. This is a diagonalization argument, adapted from Rogers' seminal textbook [23, pp. 10–11]. In [29] it is used to motivate dialetheic machines.

Claim: *There are properly paraconsistent computiations.*

Consider functions taking natural numbers as inputs and giving natural numbers as outputs. An algorithm is a procedure for computing a function (a device with a unique output); *algorithmic functions* are, intuitively, effectively computable functions. An algorithmic function can be formalized as a finite string of symbols, and it is decidable whether or not a string constitutes an algorithmic function. So there is an enumerable, comprehensive list: all the strings of length 1; all the strings of length 2; and so forth. The list itself is infinite, but each member of the list is reached by some finite stage. Consider the list $\mathbb{A}$ for algorithmic functions in just one variable. Let F_x be the $(x+1)^{th}$ member of $\mathbb{A}$, and f_x be the corresponding function. Define

$$\eth(x) := f_x(x) + 1$$

To compute $\eth$, generate $\mathbb{A}$ up to F_x, then compute f_x and add 1. This process is foolproof: therefore $\eth$ is algorithmic. Since $\eth$ is an algorithmic function in one variable, there is a z such that $\eth = f_z$ corresponding to some F_z on $\mathbb{A}$. Then

$$f_z(z) = \eth(z) = f_z(z) + 1$$

and so $n = n + 1$ for some natural number n.

Therefore $\eth$ is an inconsistent computation. To phrase it as a reductio, we could say that the comprehensive list of all algorithmic functions is not comprehensive (even though it apparently is). Or we could say that $\eth$ is not algorithmic (even though we independently argued that it is, by *giving an algorithm* to compute it)—it is both algorithmic and not. Sylvan and Copeland conclude that "there are more algorithmic functions than all algorithmic functions" [29, p. 195].

Rogers agrees that this presents itself as a serious problem [23, p. 11]; assuming consistency, it could be a disaster for recursion theory before it even gets started:

> The diagonal method would seem to throw our whole search for a formal characterization [of algorithm] into doubt. It suggests the possibility that no single formally characterizable class can correspond exactly to the informal notion of algorithmic function.

The standard solution, as Rogers goes on to say, is incompleteness. We learn the object f_z is only a *partial* function, not defined for all values (such as z itself). As with dialetheic reaction to the other logical paradoxes [16, 24], it looks like there is nothing to explain why some functions cannot take all inputs from the very domain they are meant to draw on—nothing except the inconsistency that results. Rogers draws the line thus:

> We might perversely hope to reinstate diagonalization by effectively selecting just those sets of instructions which do yield total functions...[but] if we are to avoid diagonalization, it must be the case that no algorithm for such a selection process can exist [23, p. 12].

A consequence of the incompleteness solution, then, is that no algorithm can effectively select just those sets of instructions that yield total functions. So, if consistency is a constraint, then there is terminal uncomputability in the foundations of computation. The distinction between partial and total functions, the very crux of the standard solution, cannot be effectively drawn; and there is nothing except pain-of-inconsistency to explain why not. Contrapositively, if it is possible that an algorithm for such a selection process exists, then diagonalization is unavoidable. Once the impossibility of contradiction is reconsidered, the possibility of such a procedure reopens.

2.2 The Paradox of Naive Proof

A more concrete motivation is from mathematical practice. The idea is that *proofs* are step-by-step effective procedures, that can also generate inconsistency [5, 16, 25]. We present it here in thumbnail, as a motivation for paraconsistent computation, not a definitive argument.

Naive proof is proposed as a formal, axiomatic reconstruction of proofs as they are given in mathematical English [19, p. 237]. Naive proofs are "chunks of discourse" that "amount to a compelling proof" [27], and they are treated, in practice, as bona fide proofs.[2] Azzouni describes proofs as corresponding to "derivations in one or another algorithmic system...[which] are (in principle) mechanically recognizable" [3, p. 83]. The putative upshot is that, because the mathematical community can always reach consensus about the soundness of a proof, there is an *effective* check, a computable way of establishing that a proof is valid [19, p. 41]. In a slogan, naive proof is recursive.

The claim that naive proofs are mechanically recognizable is, no doubt, contentious [21, 30]. Formalizing a proof for the purpose of checking it is fraught and, as with any translation between languages, is as much poetry as it is precision. But let us set this aside for now and follow out the implications of mechanical naive proof, to a problem. To put it colorfully, if proofs are all mechanical, then you have a *Turing mirror*—a machine that formally mimics all your proofs. This mirror is itself a mathematical object. You can reason *about* it, and it has to mimic your reasoning. What happens when you consider well-formed sentences like 'my mirror will conclude that this sentence is false'?

[2] Augmented mathematical English "could have its syntax tidied up so that it was a formal language, and the set of naive theorems expressed in this language would be deductively closed" [19, p. 41].

For the duration, let $\Vdash$ be the naive proof relation. A naive prover will be able to encode various facts about her own proving habits, via a suitable coding system, $\ulcorner\cdot\urcorner$. It seems reasonable to hope that she can ensure proofs are *representable*[3]: there is a two place arithmetic predicate $Pr(x, y)$ such that

- m is the code of a proof of sentence with code n iff $\Vdash Pr(m, n)$
- m is not the code of a proof of sentence with code n iff $\Vdash \neg Pr(m, n)$

Define a *provability* predicate: $\textsc{Prov}\ulcorner A\urcorner := \exists y Pr(y, n)$ with $n = \ulcorner A\urcorner$.

What then can the naive prover prove about her own proof relation? There is at least one basic adequacy condition to check: *soundness*. This is the uncontentious observation that, if a theorem is proven, it is ipso facto true. Proof is a source of truth. And since naive proof works with a fixed alphabet on finite strings, arguing soundness to a theorem is straightforward induction: either A is an axiom, and therefore true, or it follows from true sentences by valid rules, and so is true,

$$\textsc{Prov}\ulcorner A\urcorner \Vdash A$$

And then a corollary

$$\text{If } \Vdash A \text{ then } \Vdash \textsc{Prov}\ulcorner A\urcorner$$

since the proof of soundness itself shows that, if $\Vdash A$, then it is either a theorem or an axiom—provable either way: "if something is naively proved, then this fact itself constitutes a proof that [it] is provable" [19, p. 238]. Internalizing this claim amounts to 'reflexivity': $\textsc{Prov}\ulcorner A\urcorner \Vdash \textsc{Prov}(\ulcorner \textsc{Prov}(\ulcorner A\urcorner)\urcorner)$.

The point of all this is that naive proof thus appears to satisfy the conditions for Gödel's theorem: naive proof is a sound formal system with a recursive proof relation, able to represent recursive functions.[4] Therefore the diagonal lemma kicks in: there is a sentence G that (provably) says 'G is not provable'. Reasoning with excluded middle, if G is provable then it is true, and so not provable; so G is not provable. But that is what G says, so we've just proved

$$\Vdash G \wedge \neg G$$

As Horsten nicely summarizes the situation, "the informal notion of provability is reflexive, whereas formal [consistent] notions of provability are not" [12, p. 21]. Friends of consistency will have to explain why some apparently effective procedures (like the inductive argument proving soundness) are somehow not, and more generally, why apparently effective naive proof methods must always somehow out-

[3]This would follow by showing that all recursive functions are representable, and the (naive) proof relation is recursive. For the classical steps of this highly-non-trivial exercise, see [28], esp. Chaps. 11–12. Cf. [19, Chap. 17], [6, p. 81].

[4]In standard frameworks, soundness would imply consistency, and the more usual statement of Gödel's theorem ("no sufficiently rich consistent system..."). These are also classically equivalent to non-triviality. Here the notions come apart. See [13].

strip machines.[5] This is, more than any Liar-paradox, the centerpiece of Priest's original argument for dialetheism [16]: naive proof is inconsistent.

3 Inconsistent Computation: The Very Idea

These two paradoxes urge us to reconsider the notion of computability, widening the class of all algorithmic functions to paraconsistent objects that do not classically exist. There is much more that could be said, pushing back on these arguments, e.g. [27, 30]. But supposing that we are sufficiently interested in looking for these properly-non-classical algorithms, more practical questions arise: what is the main approach to take, and what immediate obstacles does it face?

3.1 Non-Determinisitic Machines

Agudelo and Carnielli have produced a model of computation [1], using the paraconsistent logic `LFI1*`. The main idea is to take an axiomatization of deterministic Turing machines, and apply it to non-deterministic Turing machines (NDTMs).[6] The results are called (entangled) paraconsistent Turing machines. A paraconsistent Turing machine is a NDTM such that, "when the machine reaches an ambiguous configuration, it *simultaneously* executes all possible instructions" [1, p. 580]. Agudelo and Carnielli investigate this as a simulation of phenomenon from quantum computation.

It is not too hard to see that NDTMs are equivalent in computational power to TMs: a NDTM is like several deterministic Turing machines running at once. Similarly, quantum computation is known not to surpass the power of standard Turing machines. Agudelo and Carnielli are explicit that their aims are conservative: they do not aim to break the Church-Turing thesis or extend the class of computable functions. They express a (healthy) skepticism about more radical approaches [1, p. 574, esp. footnote 2]. Their paper is a framework for modeling quantum and parallel processing.

Our introductory comments point in more radical directions. Without explicitly taking a stand on the status of the Church Turing thesis (cf. [9], [28, Chap. 34]), we are interested in the set of 'naively' computable functions—perhaps itself an inconsistent class. If the goal is to get at some mathematically novel objects, then a more unorthodox approach is required.

---

[5]Of course, friends of consistency do have answers to these questions, e.g. [28, Chap. 28]. For a good general discussion, see [26].

[6]For standard background on Turing machines, [28, Chaps. 31–33] is reccomended. For other paraconsistent approaches, the idea of non-deterministic matrices for paraconsistent logic has been studied by Avron, Zamansky, et al. [2]. Since truth tables are rudimentary computers, one can take this work to be in a similar vein; cf. [33].

3.2 Dialetheic Machines

Sylvan and Copeland sketch the full dialetheic plan [29, pp. 197–198]: to treat computation in an inconsistent metalanguage. I quote their 'dialethic machines' passage (with Sylvan's preferred spelling 'dialethism') at length:

> It is not difficult to describe how a machine might encounter a contradiction: for some statement A, both A and $\sim A$ appear in its output or among its inputs. ... [A] machine programmed with a dialethic logic can proceed with its computation satisfactorily. Let us call such machines D-machines of type 1 (D for *Dialethic*). D-machines of type 2 are machines whose *meta* logic is dialethic: for such a machine, M, one of whose states is x, 'M is in x' and '$\sim$(M is in x)' may both be the case. D-machines of type 1 are nothing more than Turing machines, albeit a conspicuously useful sort of Turing machine if inconsistent data is in the offing. .... [A] central idea of paraconsistent computability theory: such Turing machines may be employed to compute diagonal functions that are classically regarded as uncomputable. It is an open question whether D-machines of type 2 compute classically uncomputable functions and, if so, which. We recommend the question to the paraconsistent community.

From the text we can discern at least two open problems about dialetheic machines: (1) Do they break the Church-Turing barrier? (2) What is that status of the halting problem?

Taking up the call from Sylvan and Copeland afresh,[7] our interest is in the abstract, mathematical objects of dialetheic machines—the theoretical software, so to speak, for D-machines of type 2. The aim from here, in keeping with the motivations, is to get a feel for what it would be like to work with no classical metatheory whatsoever—looking for unreconstructed inconsistency, so to speak. As far as hardware questions, we do at the end of Sect. 3.3 below try to address what a machine M of which it is true to say 'M is in state x and also not in state x' might look like; cf. [33]. The discussion, though, remains at the theoretical or 'pure' level.

3.3 Inconsistent Proofs

Once one is thinking about dialetheic machines, the main conceptual hurdle to clear is the natural question: when we actually turn the machine on and press 'GO', *what does it do?*

To make things more concrete, we return to our competent, diligent clerk who is capable of carrying out algorithms, and imagine her carrying out proofs within true mathematical theories. We can see how to address dialetheic machines in general by trying to think about paraconsistent proofs.

[7] Agudelo and Carnielli's ParTMs yield negative answers to both questions. However, even granting their assertion that ParTMs are D-machines of type 2, a negative answer for them does not exhaust the space of possible solutions. And so while ParTMs lead neither to hypercomputation nor to revising the halting problem, this does not constitute a complete or definitive answer to the open questions.

Already in Sect. 2.2 above we had the naive derivation of a contradiction, the 'Gödel paradox' that $\Vdash G \wedge \neg G$. In a critique of dialetheism, Stewart Shapiro has observed that these internal inconsistency facts ought to make their way out to the metalevel [27]; cf. [19, pp. 239–243]. The internal inconsistency is

$$\Vdash \exists x Pr(x, \ulcorner G \urcorner) \quad \text{and} \quad \Vdash \neg \exists x Pr(x, \ulcorner G \urcorner)$$

A fortiori, by soundness, G is provable and not provable. But now the extra step: if the biconditionals in the representation of proof contrapose,

- m is not the code of a proof of sentence with code n iff $\nVdash Pr(m, n)$
- m is the code of a proof of sentence with code n iff $\nVdash \neg Pr(m, n)$

then the negation facts on the Pr predicate push out to the relation $\Vdash$. The conditions on proof itself take on the appearance of 'consistency'

$$\Vdash Pr(m, n) \quad \text{iff} \quad \nVdash \neg Pr(m, n)$$
$$\Vdash \neg Pr(m, n) \quad \text{iff} \quad \nVdash Pr(m, n)$$

which (ironically?) generate a stronger form of contradiction:

$$\Vdash G \quad \text{and} \quad \nVdash G$$

This is not merely a contradiction *in* naive proof theory; this is a contradiction *about* naive proof theory.

Aside from suggesting that, through Gödel coding, there is an inconsistency about the natural numbers (!), this suggests that there is an inconsistent proof (!!). What is an inconsistent proof? As Shapiro puts it [27, p. 828–9, some symbolism changed],

> On all accounts—including the non-dialetheic perspective—we have that n is the code of a derivation of G. This can be verified with a painstaking but completely effective check. How can the dialetheist go on to maintain that, in addition, n is not the code of a derivation of G? What does it mean to say this? Since $\neg Pr$ is a recursive predicate, we can supposedly verify—at the same time, in almost the exact same way—that n is *not* the code of a derivation of G. How? ... All goes well everywhere, and something goes wrong somewhere.... Some *one* step in the procedure must yield contradictory results. Which step can do that?

The questions are largely rhetorical, I take it. Here is an answer.

As a general point, the 'all well everywhere/something wrong somewhere' dynamic is a very apt description for dialetheic data. In truth theory, all contradictions are false; some of them are also true. In vagueness, all possible cutoff points for the predicate 'is a heap' are unbelievable; some cutoff point must also be correct, since everything is not a heap. In set theory, all ordinal numbers $\alpha = \{\beta \in On : \beta < \alpha\}$ are strictly ordered, $\alpha \nless \alpha$; also, some ordinal number $On < On$. And so forth. So qua dialetheism, there is nothing immediately special to say about Shapiro's objection.

What may seem extra peculiar is that a proof is a finite object, open to exhaustive inspection, unlike a transfinite ordinal. But a heap of sand is finite, too (for the record).

What needs explaining, then, is how some specific step in a proof can be *both valid and invalid*. The answer may be more banal than one might have expected. A proof of G looks like this:

$$\pi_G = \langle A_0, \ldots, A_n, G \rangle$$

Now, as we tell students, a proof is invalid if it can have premises true but conclusion false. With this intuition, then of course π_G is not a proof of G, because it is invalid. The premises $A_0, \ldots, A_n$ are axioms or theorems of arithmetic, and therefore true; but $\neg G$ is a theorem, so G is false. Shapiro asks, which step was invalid? The answer is: the step from truth to falsity, which at the very latest occurs at G. Any proof of a contradiction is also not a proof, because all contradictions are false.

The guiding intuition, then, is that what a dialetheic machine does when you turn it on is...what any other machine would do. Some *descriptions* of its actions will be inconsistent, but all the individual actions are exactly as they've always seemed, to dialetheists and non-dialetheists alike. This is only so much story-telling, though. Let us move to some more details.

4 Implementing Inconsistent Computation

The foregoing amounts to an abstract prediction of a legitimately inconsistent algorithmic procedure. The dialetheic machinist has work to do.

4.1 On Whether $0 = 1$

In the paradox of all algorithmic functions Sect. 2.1, on assumption, if diagonal $\mathfrak{d}$ is algorithmic then $0 = 1$; so by standard reductio, it is not a algorithmic. We wish to say that $\mathfrak{d}$ is algorithmic: there is a clear recipe to compute it. To maintain this conclusion, there are at least two ways to do it:

- Accept $0 = 1$

This might sound like madness, but it can be done, in at least two ways. (Cf. [19, Chap. 18].) First, one could arrange that $0 = 1$ can hold without absurdity, $\bot$, by using a weak definition like $0 := \{x : x \neq x\}$. This 'zero' need not be empty; it contains the Russell set, for example. Second, one could even allow $0 = 1 \supset \bot$ and $0 = 1$ both to hold, but have $\supset$ be the material conditional and so fail to obey modus ponens.[8] Sylvan and Copeland seem to indicate that accepting $0 = 1$ is tenable.

There are some severe drawbacks for this sort of approach. For the first case, defining zero weakly makes it not behave much like 0 [19, p. 253]. In the second case, using $\supset$, we lose modus ponens. But in any case, I am not rushing to clear brush on this path. If $0 = 1$ is acceptable, then the original 'paradox of all algorithms'

---

[8]The material conditional is $p \supset q := \neg p \vee q$, so 'modus ponens' for it $p, p \supset q \therefore q$ is just disjunctive syllogism, which is not paraconsistently valid. See [19, Chap. 8].

reductio does not go through! Such an arithmetic base is so weak, it has defused its own *raison d'être*. There are no algorithmic functions that are also not algorithmic; there is at most an inert contradiction in arithmetic.

This makes the other, more attractive option, more attractive:

- Avoid $0 = 1$

How can this be done? By rebuilding recursive function theory in an appropriate inconsistent arithmetic, with non-self-identical programs, so that a properly inconsistent simulacra of $\eth$ can be built, but one which does not churn out pure noise.

This is a rather more involved project, to put it mildly. For this chapter, we will leave the further technical options as open as possible.[9] In this vein, the expression is informal, e.g. we use the natural language 'if...then...', 'implies', 'iff' etc. on the assumption that it may be specified later as a paraconsistent conditional, in any number of suitable systems. Minimally, from the discussion above we presume that any conditional will satisfy at least some minimal conditions (p implies p, transitivity) and obey modus ponens; it will also help to have a contraposable conditional, but we will flag any place that contraposition is appealed to. The law of excluded middle and arguments by cases are taken for granted.

4.2 Functions

The most immediate problem for founding a 'theory of paraconsistent functions' is that functions (as usually conceived) do not perform at all well with inconsistency. The characteristic function of the Pr predicate is

$$g_{Pr}(x, y) = \begin{cases} 1 & \text{if } Pr(x, y); \\ 0 & \text{if } \neg Pr(x, y) \end{cases}$$

But then, given the Gödel sentence, it follows that $1 = g_{Pr}(\pi_G, G) = 0$.

What are we to say about this? Either $0 = 1$, or identity is not transitive, or g is not a classical function. The first case has been dealt with; the second case is explored in [20] but again requires a non-ponenable $\supset$ to work. Let us investigate the tenability of the third case.

Going forward, we assume that there is a set $\mathbb{N}$ called 'the natural numbers' (see [14, 17]), and stipulate that the pair

$$\{0, 1\} := \{x : \text{if } x \neq 0 \text{ then } x = 1\}$$

is non-empty. For a material 'if...then...', this is equivalent to the more familiar $\{x : x = 0 \vee x = 1\}$, but without disjunctive syllogism, the conditional phrasing is

[9] The background is assumed to be a paraconsistent set theory/arithmetic, along the lines of [7, 24], but the foundational details are not the main issue here. Readers may take assertions of the existence of e.g. relations as accomplished by axiomatic fiat, with foundations to be filled in elsewhere.

preferred.[10] We do get, by excluded middle, that if $x \in \{0, 1\}$ then $x = 0$ or $x = 1$, but not vice versa. With contraposition, the definition gives $\{0, 1\} = \{x : \text{if } x \neq 1 \text{ then } x = 0\}$, too. And modulo the discussion of $0 = 1$ above, we have

$$0 = 1 \text{ implies} \bot$$

where $\bot$ implies any sentence whatsoever. These modest requirements will generate some 'functional' properties below.

Let g be a relation $g : \mathbb{N} \longrightarrow \{0, 1\}$. Some notation: for two place relations taking natural numbers to $\{0, 1\}$,

$$\langle a, b\rangle \in g \subseteq \mathbb{N} \times \{0, 1\}$$

we can write the usual $b \in g(a)$, as well as the more suggestive

$$g(a) \ni b$$

This can be pronounced '$g(a)$ outputs b'. For any A, its *characteristic relation* g_A is

$$g_A(x) \ni \begin{cases} 1 & \text{iff } A(x); \\ 0 & \text{iff } \neg A(x) \end{cases}$$

Characteristic relations still track whether or not x is A; they are entirely deterministic; they just allow overdetermination: the characteristic relation of the Pr predicate is

$$g_{Pr}(x, y) \ni \begin{cases} 1 & \text{iff } Pr(x, y); \\ 0 & \text{iff } \neg Pr(x, y) \end{cases}$$

Then $1 \in g_{Pr}(\pi_G, G) \ni 0$.

In many cases, characteristic relations are indistinguishable from functions, especially if we make explicit the negation clauses (which can either be stipulated, or follow automatically if the 'iff' is contraposable):

$$g_A(x) \not\ni \begin{cases} 1 & \text{iff } \neg A(x); \\ 0 & \text{iff } A(x) \end{cases}$$

Any characteristic relation then has 'function' properties:

1. **Exclusive**: $g(a) \ni 1$ iff $g(a) \not\ni 0$; $g(a) \not\ni 1$ iff $g(a) \ni 0$
2. **Discriminating**: If $b \neq c$ and $g(a) \ni b$ then $g(a) \not\ni c$
3. **Materially univocal**: If $g(a) \ni b$ then $g(a) \not\ni c$ or $b = c$

[10]In this we are following Dunn's axiomatization of relevant robinson arithmetic [11], where he explains that this phrasing is "less deductively sterile".

EXCLUSIVITY is from putting together the positive and negative defining clauses of characteristic relations. Then relations are DISCRIMINATING as follows. Let $b, c \in \{0, 1\}$, so they are both either 0 or 1. Assume $b \neq c$. There are then four cases to consider, or only two without loss of generality. If $b = 0$ and $c = 1$, that's good; if $b = 0 = c$, then ex hypothesis $0 \neq 0$, and then using the conditionalized definition of $\{0, 1\}$, we get $0 = 1$ which implies $\bot$. (Mutatis mutandis for $b = 1 = c$.) So either way, the premise $b \neq c$ ensures that exactly one of b and c are 0 and exactly one of them is 1. Then using EXCLUSIVITY, if $g(a) \ni 1$ then $g(a) \not\ni 0$ and vice versa. Generalization gives the result. Finally, MATERIAL UNIVOCALITY follows by contraposing on DISCRIMINATING, or else repeating the reasoning by cases in a slightly different order.

It is worth emphasizing that (3), MATERIAL UNIVOCALITY, is *classically equivalent to the classical definition* of function. We have not merely replaced a functional input-output Turing machine with a non-functional relation. We have familiar machines, but where unfamiliar assumptions must be articulated: g is a function—or else there is something it both outputs and also doesn't. The latter clause is usually simply never considered, but it is always there. So any failures of functionality are accounted for with the additional possible case of inconsistency—inconsistency which is generated by insisting on extra 'consistency' properties, in EXCLUSIVITY and the definition of $\{0, 1\}$. This all generates an object which to the classical eye is rightly called a function.

In the event of some contradictory $A(a) \wedge \neg A(a)$, then the relation will report $g_A(a) \ni 1$ and $g_A(a) \not\ni 1$. Then

4. **Inconsistency**: If $\langle a, 1\rangle \in g$ and $\langle a, 1\rangle \notin g$ then $g \neq g$

since by set extensionality g differs from itself with respect to membership, and so is not identical to itself. Therefore as conceived here, a paraconsistent function does not merely process inconsistency, but is *itself* inconsistent: in the language of Sect. 3.2, a type-2 dialetheic machine.

4.3 The Halting Problem

How does this play out against the famous halting problem?[11] In the standard picture of computation, some programs halt on input, and some do not. Computing whether or not a program π halts on an input is to consider two options; in a flow-chart,

[11] See [28, Chap. 33] for background.

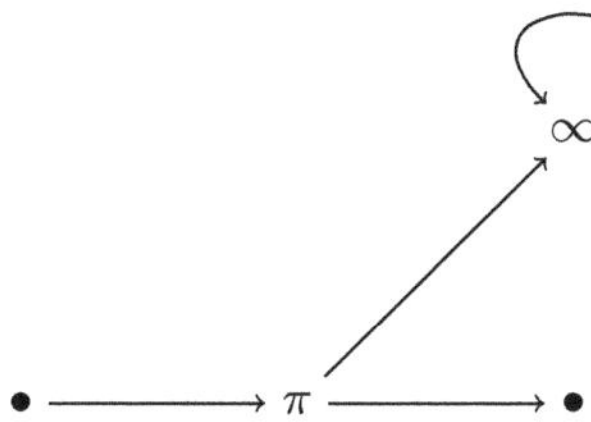

depicting an input that either sends π into some infinite routine, or else arrives at an output and stops. As was remarked in the early days of recursion theory, it would be good to have an effective check for which routines terminate, and which never come back. To capture this with notation, for any f, if f does not terminate on n, as a kind of slang we'll write $f(n) \ni \infty$, and if it does terminate then $f(n) \not\ni \infty$.

Consider a relation $\mathfrak{halt} : \mathbb{N} \times \mathbb{N} \longrightarrow \{0, 1\}$ such that

$$\mathfrak{halt}(x, y) \ni \begin{cases} 1 & \text{iff } x(y) \not\ni \infty; \\ 0 & \text{iff } x(y) \ni \infty \end{cases}$$

Then $\mathfrak{halt}$ can be used to define $\mathfrak{d}$ as follows[12]:

$$\mathfrak{d}(x) \ni \begin{cases} 1 & \text{iff } \mathfrak{halt}(x, x) \ni 0; \\ \infty & \text{iff } \mathfrak{halt}(x, x) \ni 1 \end{cases}$$

Therefore

$$\mathfrak{halt}(\ulcorner \mathfrak{d} \urcorner, \ulcorner \mathfrak{d} \urcorner) \ni 1 \text{ iff } \mathfrak{d}(\ulcorner \mathfrak{d} \urcorner) \ni \infty \text{ iff } \mathfrak{halt}(\ulcorner \mathfrak{d} \urcorner, \ulcorner \mathfrak{d} \urcorner) \ni 0$$

With the law of excluded middle, or at least a reductio principle, these biconditionals yield contradictions:

$$\mathfrak{halt}(\ulcorner \mathfrak{d} \urcorner, \ulcorner \mathfrak{d} \urcorner) \ni 1 \quad \text{and} \quad \mathfrak{halt}(\ulcorner \mathfrak{d} \urcorner, \ulcorner \mathfrak{d} \urcorner) \ni 0$$

and

$$\mathfrak{halt}(\ulcorner \mathfrak{d} \urcorner, \ulcorner \mathfrak{d} \urcorner) \not\ni 1 \quad \text{and} \quad \mathfrak{halt}(\ulcorner \mathfrak{d} \urcorner, \ulcorner \mathfrak{d} \urcorner) \not\ni 0$$

By extensionality, $\mathfrak{halt} \neq \mathfrak{halt}$.

More intriguingly, for the diagonal, we have $\mathfrak{d}(\ulcorner \mathfrak{d} \urcorner) \ni \infty$ and $\mathfrak{d}(\ulcorner \mathfrak{d} \urcorner) \not\ni \infty$. The diagonal halts and does not halt. In a final section, we will try to say more about what this could mean.

---

[12] See [23, p. 24]; the '∞' case can be specified by any process that will go on forever, e.g. 'lather, rinse, repeat'. This $\mathfrak{d}$ is not a characteristic relation onto $\{0, 1\}$ as defined above, but the only 'functional' property needed to make this argument go is EXCLUSIVITY, which is stipulated anyway; all that matters is that it is a mathematical object that exists if $\mathfrak{halt}$ does.

4.4 Finite and Infinite Machines

The lines and nodes in the diagram above are pictures of objects that ultimately exist in arithmetic. If the background arithmetic is classical PA, then we know a lot about it. What about in paraconsistent arithmetic? In particular, consider *collapsed models* of arithmetic [14, 15, 18]. These are *finite* models of numerical succession that can look like this (with these arrows representing successor, not the flow-chart arrows in the previous diagram):

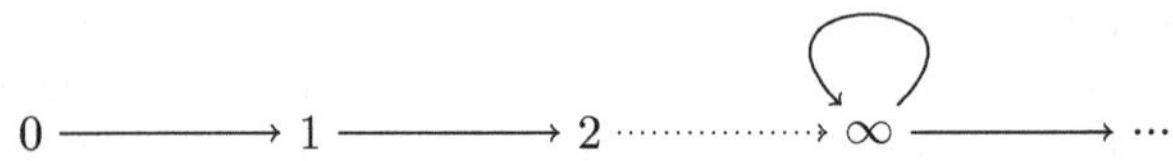

The idea is to define an equivalence relation on the natural numbers, and then treat as identical any two numbers that are so equivalent. The first number at which such a collapse occurs is denoted ∞ above.[13] The interest of these collapsed models comes from the fact that in them, no truths are lost: if a sentence is satisfied in a standard model of arithmetic, it remains satisfied in a collapsed model [19, p. 232]. And there is a salient number, $\infty = \infty + 1$.

Most logicians who think about such things have regarded these models as pragmatic ways to prove non-triviality; dialetheists and finitists have taken them more seriously, seeing in ∞ either a least inconsistent number [17] or the last finite number [32]. In either interpretation, after ∞ everything is settled: for all $n > \infty$, the model has it that $n = \infty$, so there is nothing more to discover in continuing out the successor line (other than perhaps adding the negations of established facts). In the notation of the previous section, $f(n) \ni \infty$ meant that f does not halt on n; now we read it, f gets all the way to 'the end of the numbers', which can be charitably taken an oblique way of saying what is usually meant by 'not halting'. But since ∞ is a number, if f goes on 'forever' to ∞ then f also halts. That is, if we take inconsistent arithmetic *very* seriously, we get a flow-chart of computation

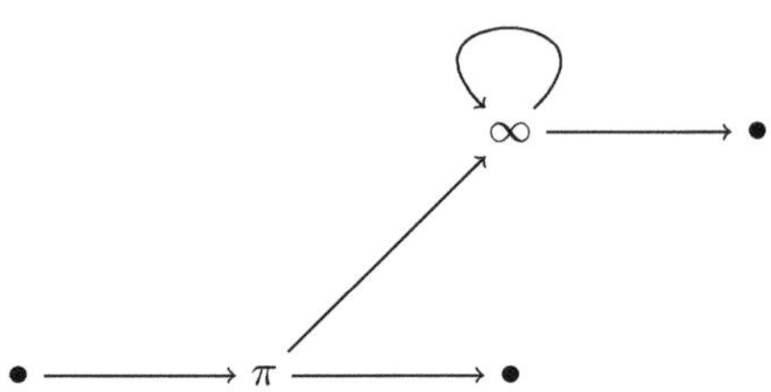

[13] Succession on this conception is not a classical function, since clearly there is at least one node out of which are coming two separate arrows. There are mainly unanswered questions in the philosophy of inconsistent mathematics: how 'long' is the cycle between a number and itself? how many numbers come after the first inconsistent number? what does 'finite' or 'infinite' mean here? etc. We must defer to another day.

where as before, either a program halts on input, or it goes off into an infinite subroutine—but on that second path, after the process has gone on for ∞ many steps, it reaches the last one, has no more steps to take. That branch of the flowchart ends too. While there are many details to fill in here (if one so desired) the basic idea is simple enough. Since both possibilities on input are 'finite', all computations halt.

It is not my intention to defend finitism; see [32], about which some questions would be easier to answer than others. For example, what about a simple program that oscillates between 0 and 1; at ∞ is the output 0 or 1? The answer would seem to be: if ∞ is even, then 0, and if odd than 1. How to conceptualize these things is more difficult. E.g. after ∞ steps, perhaps we should say that the internal state of the machine is no longer sensibly describable, but distinguish this from the computation *halting* at this step.[14] Or perhaps we should say that when a phenomenon become inconsistent, even unstable, it can still be "sensibly describable", as dialetheic. This turns on what counts as sensible, about which I shall recuse myself for now.

It should be pointed out that the extreme conclusions of this section are of limited interest, since it is a consequence of a sort of finitism, more than anything paraconsistent/dialetheic. If finitism is true, then the halting problem is trivial: the answer is always 'yes'.[15] On the other hand, if the idea is not infinitistic but inconsistent, then still no hints have been given about where on the number line ∞ lies. If it is an extraordinarily large number, something well beyond the number of subatomic particles in the universe, with "no physical meaning or psychological reality" [17, p. 338], then in terms of *feasibility* the halting problem remains a problem. Alternately, if ∞ is small enough, then the halting function is *constant*,[16] and the problem is trivialized again—but now for even more discomfiting reasons.

5 Rise of the Inconsistent Machines

In looking for novel inconsistent computations, this chapter has been about, in effect, how diagonal arguments in the language of Turing machines translate into dialetheism. A patient reader may want to interject, though, that the hard core of the halting problem, the paradox of all algorithms, etc., is not the mere derivation of a Russellesque contradiction. The problem with these diagonal arguments is that they show that computations do not always work as expected: the aim of $\mathfrak{halt}$ is to *predict* the behavior of all programs, but $\mathfrak{halt}$ must get the diagonal *wrong*! Listen to Tarski:

> We know (if only intuitively) that an inconsistent theory must contain false sentences; and we are not inclined to regard as acceptable any theory which has been shown to contain such sentences [31].

---

[14]Thanks to referees here.

[15]Similarly, by the Church-Turing thesis, the unsolvability of the halting problem implies the undecidability of first order logic. I do not suggest otherwise. I am just noticing that if the above picture of arithmetic were accurate, then first order logic is automatically decidable.

[16]Thanks to Tomasz Kowalski for pointing this out.

Isn't that the real sticking point for any would-be 'naive computability theory'?

Sensible as these sentiments are, Tarski is only reporting half the story. We also know (to a certainty) that consistent theories leave out true sentences; and we are not inclined to regard as acceptable any theory which has been shown to leave out such sentences. The choice here is: be falsity-avoiding at all costs, and so accept incompleteness; or be truth-seeking at all costs, and so accept some inconsistency. What it means to have a dialetheic machine is that some sound computations are *not* impossible, but their outputs are false. The halting program can get some computations wrong—as long as it gets everything right. If we want to reckon with *all* computations, there will be surprises, but there is a way.

Thanks

Audiences at the Paraconsistent Reasoning in Science and Mathematics conference, Münich, the University of Aberdeen, the Melbourne Logic Group, the Otago Logic Group, and the Czech Academy of Sciences. Thanks in particular to Toby Meadows for detailed comments on an earlier draft, and two anonymous referees. Research was funded by the Marsden Fund, Royal Society of New Zealand.

References

1. Agudelo, J., & Carnielli, W. (2010). Paraconsistent machines and their relation to quantum computing. *Journal of Logic and Computation, 20*(2), 573–595.
2. Avron. A., & Zamansky, A. (2011). Non-deterministic semantics for logical systems–a survey. In *Handbook of Philosophical Logic, vol. 16*, p. 227304. Berlin: Kluwer Academic Publishers.
3. Azzouni, J. (2004). The derivation-indicator view of mathematical practice. *Philosophia Mathematica, 3*(12), 81–105.
4. Batens, D., Mortensen, C., Priest, G., & van Bendegem, J-P. (Eds.), (2000). *Frontiers of Paraconsistent Logic*. Philadelphia: Research Studies Press.
5. Berto, F. (2009). The gödel paradox and wittgenstein's reasons. *Philosophia Mathematica, 12*(2), 208–219.
6. Berto, F. (2009). *There's Something About Gödel*. New Jersey: Wiley-Blackwell.
7. Brady, R. (2006). *Universal Logic*. Stanford: CSLI.
8. Cooper, S.B., Lowe, B., & Sorbi, A. (Ed.). (2008). *New Computational Paradigms: Changing Conceptions of what is Computable*. Berlin: Springer.
9. Copeland, B. J., & Sylvan, R. (1999). Beyond the universal turing machine. *Australasian Journal of Philosophy, 77*, 46–66.
10. Da Costa, N. Paraconsistent mathematics. In Batens et al. [4], pp. 165–180.
11. Michael, J. (1980). *Dunn. Relevant robinson's arithmetic. Studia Logica, 38*, 407–418.
12. Horsten, L. (2005). Remarks on the content and extension of the notion of provability. *Logique et Analyse, 189–192*, 15–32.
13. Meyer, R.K. (1976). Proving completeness relevantly for r. typescript.
14. Meyer, R. K., & Mortensen, C. (1984). Inconsistent models for relevant arithmetics. *Journal of Symbolic Logic, 49*, 917–929.
15. Mortensen, C. (1995). *Inconsistent Mathematics*. Dordrecht; New York: Kluwer Academic Publishers.
16. Priest, G. (1979). The logic of paradox. *Journal of Philosophical Logic, 8*, 219–241.
17. Graham, P. (1994). Is arithmetic consistent? *Mind, 103*(411), 337–349.

18. Priest, G. (1997). Inconsistent models of arithmetic part i: Finite models. *Journal of Philosophical Logic*, *26*, 223–235.
19. Priest, G. (2006). *In Contradiction: A Study of the Transconsistent*. 2nd edn. Oxford: Oxford University Press.
20. Priest, G. (2014). *One*. Oxford: Oxford University Press.
21. Rav, Y. (1999). Why do we prove theorems? *Philosophia Mathematica*, *7*, 5–41.
22. Restall, G. (1992). A note on naïve set theory in LP. *Notre Dame Journal of Formal Logic*, *33*, 422–432.
23. Rogers, H. (1967). *Theory of recursive functions and effective computability*. McGraw-Hill series in higher mathematics: McGraw-Hill.
24. Routley, R. (1977). Ultralogic as universal? *Relevance Logic Newsletter*, *2*, 51–89.
25. Routley, R. (1979). Dialectical logic, semantics and metamathematics. *Erkenntnis*, *14*, 301–331.
26. Shapiro, S. (1998). Incompleteness, mechanism, and optimism. *Bulletin of Symbolic Logic*, *4*, 273–302.
27. Shapiro, S. (2002). Incompleteness and inconsistency. *Mind*, *111*, 817–832.
28. Smith, P. (2007). *An Introduction to Gödel's Theorems*. Cambridge.
29. Sylvan, R., & Copeland, J. (2000). Computability is logic relative. In D Hyde, G Priest, (Eds.), *Sociative Logics and their applications*, pp. 189–199. Ashgate.
30. Tanswell, F. (2014). Saving proof from paradox: Against the inconsistency of informal mathematics, 2014. Presented at 'Paraconsistent Reasoning in Science and Mathematics', LMU/MCMP, June 12.
31. Tarski, A. (1944). The semantic conception of truth and the foundations of semantics. *Philosophy and Phenomenological Research*, *4*, 341–376.
32. Van Bendegem, J.P. (2003). Classical arithmetic is quite unnatural. *Logic and Logical Philosophy*, *11*, 231–249.
33. Weber, Z., Badia, G., Girard. P. (2016). What is an inconsistent truth table? *Australasian Journal of Philosophy*, *94*(3), 533–548.

MIX
Papier aus verantwortungsvollen Quellen
Paper from responsible sources
FSC® C105338

If you have any concerns about our products, you can contact us on
ProductSafety@springernature.com

In case Publisher is established outside the EU, the EU authorized representative is:
Springer Nature Customer Service Center GmbH
Europaplatz 3, 69115 Heidelberg, Germany

Printed by Libri Plureos GmbH
in Hamburg, Germany